Progress in IS

"PROGRESS in IS" encompasses the various areas of Information Systems in theory and practice, presenting cutting-edge advances in the field. It is aimed especially at researchers, doctoral students, and advanced practitioners. The series features both research monographs that make substantial contributions to our state of knowledge and handbooks and other edited volumes, in which a team of experts is organized by one or more leading authorities to write individual chapters on various aspects of the topic. "PROGRESS in IS" is edited by a global team of leading IS experts. The editorial board expressly welcomes new members to this group. Individual volumes in this series are supported by a minimum of two members of the editorial board, and a code of conduct mandatory for all members of the board ensures the quality and cutting-edge nature of the titles published under this series.

Jürg Meierhofer · Shaun West · Thierry Buecheler
Editors

Smart Services Summit

Smart Services Creating Sustainability

Springer

Editors
Jürg Meierhofer
Zurich University of Applied Sciences
Winterthur, Zürich, Switzerland

Thierry Buecheler
Oracle
Zürich, Switzerland

Shaun West
Institute of Innovation and Technology Management
Lucerne University of Applied Sciences and Arts
Horw, Luzern, Switzerland

ISSN 2196-8705 ISSN 2196-8713 (electronic)
Progress in IS
ISBN 978-3-031-36697-0 ISBN 978-3-031-36698-7 (eBook)
https://doi.org/10.1007/978-3-031-36698-7

This Springer imprint is published by the registered company Springer Nature Switzerland AG
The registered company address is: Gewerbestrasse 11, 6330 Cham, Switzerland

Foreword

Smart services form an engine for value creation today. What are they? Why are they important? How can we approach them? Let us make a decomposition: smart services = services x smartness.

The first item, services, has grown over the past decades to an essential component, especially in post-industrial economies. Services provided by manufacturers often form so-called products/service systems (PSSs) or product-as-a-service (PaaS). Research and innovation of PSSs and services themselves have been active for the last decades: one major question is if PSSs and PaaS bring positive contributions to environmental impacts such as CO_2 emissions and natural resource use. Repair services on products, for instance, are expected to prolong the product lives and thus save resources used. Expecting less environmental impacts in general from PSSs has been indeed sensible to many working in industry, to policymakers and to those in academia; therefore, it has been attractive as a way for a business to aim at more sustainability while staying competitive. Although a number of scientific questions about PSSs remain unanswered, much insight has been developed and exploited in industry.

The second item, smartness, at present, often means intelligence realized through big data, connectivity and reasoning. Smart services gained momentum in industry thanks to IoT and big data analytics. As you are interested in this book, you probably agree that a service has found its element—'a fish in water' in Japanese (my native language). As a typical example, remote monitoring and analytics services for products in operation could reason about the break-down probability of a specific component during the next week and thus help avoid a more severe malfunction of the whole product or system. Intelligence is often capitalized for dynamic adaptation and thereby higher value. Using the available knowledge on PSSs or services, smart services will surely be better exploited in practice. Expanding on the earlier stated PSS question, we may want to seek answers to a high-level question: do smart services contribute to environmental sustainability? This is a highly complex issue because of the dynamics and enlarged system, just to name a few.

Looking at the contents of the summit, all the questions asked are relevant and will contribute to answering the high-level question above. The summit was organized by one of the internationally known groups in this area. The questions are all timely, just considering the heightened attention and expectations given to a circular economy. Many companies are attempting to find solutions for more circularity using smart services. Realizing the traceability of products and components throughout multiple use phases and providing the right information at the right time by smart services may be an excellent example: before a product comes to a service workshop, you could be informed that components A and B are to be replaced and other components will function over the next use phase. Technologies are there, but how to use the technologies is another critical question for sustainability. Here, systemic thinking with a systems perspective is critical. As an approach, we need multiple disciplines, and industry-academia interaction is essential: namely, so-called trans-disciplinary 2 research is highly recommended. I am sure this book will provide you with some hints about the high-level question above.

Read it and get inspired!

Linköping, Sweden
November 2020

Tomohiko Sakao

Preface

Data Innovation Alliance—Expert Group Smart Services
The data innovation alliance provides a significant contribution to making Switzerland an internationally recognized hub for data-enabled value creation. With its initiative 'NTN databooster', which is supported by Innosuisse, companies get access to a proven innovation methodology with a systematic and facilitated innovation process.

Oracle
We were fortunate this year to be sponsored by Oracle and to have our summit in their offices in the Circle at Zurich airport. The location was terrific, the food delicious, the drinks refreshing and the keynote thought-provoking. The ambience was conducive to rich and open discussions.
ZHAW and Hochschule Luzern
Both universities must be thanked for their support and encouragement that they provided to the two co-chairs of the Summit.

Zürich, Switzerland — Jürg Meierhofer
Horw, Switzerland — Shaun West
Zürich, Switzerland — Thierry Buecheler

About the Summit

As the adoption of Smart Services continues to accelerate, we see an opportunity to apply the emerging technology to provide new value propositions and business models that provide tangible sustainability outcomes that help us and the planet. Fig. 1 provides the artwork for the summit. In 2021, we held the Smart Service Summit 'in person' at a fantastic location overlooking the lake of Zurich. The focus was on the impact of COVID-19, and in hindsight, we were between two outbreaks. This year we tried to leave COVID-19 behind us and look to build a brighter future with a focus on sustainability. The summit in 2022 aims to assess new and emerging services that are enabled by technology and where the services are co-delivered to support sustainability. In doing so, we hope to answer some of these questions:

- how are the service quality and value impacted through digital technologies?
- how can you transform the customer (or a third party) into a service partner?
- how does collaborative working impact value co-creation?
- what is the impact of smart services on customer experience?
- how does the nature of the service delivery change?
- how can organizations be enabled to foster digital value co-creation?

We had over 70 authors contribute to the papers, and of those, we had 42 attend the summit. 20 academic papers were accepted from those submitted. All the academic papers were double-blind reviewed. The authors were from 10 counties, with Switzerland, Germany and Austria representing 60% of the authors, and other authors were from Italy, Canada, the UK, the USA, India, Mexico and Sweden. We had a mix of different disciplines from Design, Engineering, IT, ecology and data science. Around 30% of the contributors were practitioners. This all contributed to the diversity and supported the creativity of the group—missing was gender diversity, as the number of women who attended the summit was only 20%. The additional diversity of perspectives would have been welcomed.

Fig. 1 The Smart Service Summit 2022

Contributions

I (Shaun West) want to start with a quote from my friend Larry Liefer:

> *All products deliver a service, BUT, the design profession has only focused on the stuff, not the service.*

This quote is important as it sets the scene for smart services as it highlights what we set up the summit to focus on. It also provides a link with the word cloud based on the abstracts of the papers (Fig. 2).

This year we for fortunate enough to have five excellent keynotes at the summit (Fig. 3).

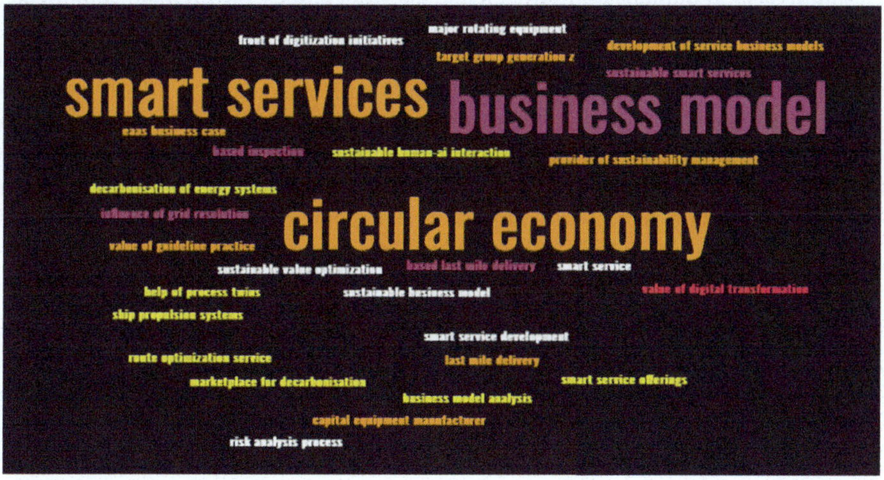

Fig. 2 Word cloud from the abstract titles

Fig. 3 Our keynote speakers

Thierry Buecheler opened the keynotes with a presentation, 'smart services for people', where he grounded the six questions for the summit around real-world cases based on experience with a hospital in the UK (Fig. 4). His review provided insights:

- **Digital**. A mixed team came up with a simple digital service to provide a second opinion from those world-leading doctors for patients all around the globe. This was only realistically achievable through digital channels.
- **Partners**. The change was only possible by jointly designing and owning the end-to-end process, including domain experts beyond 'digital'.
- **Co-creation**. Creating the 'perfect service' is nice. However, if it is not understood and 'owned' by the involved parties, and nobody will use it in an intended way, it is useless. Imagine an AI black box (e.g., a Neural Network most people don't grasp intuitively). If humans don't trust it and get what it does, or if it's too complicated to use, it will not be effective—independent of good service design.
- **Delivery**. MUCH faster & MUCH more volume. Non-linear service usage (humans tend not to follow processes/protocol. Also, they tend not to act or decide rationally. Digital might give them more choices. They tend to use multiple channels, devices, sources, opinions in their respective bubbles ...)
- **Customer experience**. The design of the service might be perfect, but if the delivery doesn't work, it's useless. In the context of customer (or patient) experience, this includes GUIs and workflows around the service(s).

Alain Schilli provided the second keynote on 'smart services for sustainability/circular economy', in which he focused on learning from nature. Alain was inspired and educated by nature, observing living matters and trying to understand what matters, and this is a guiding approach that he uses in the transition from an old economy to a sustainable new economy. To achieve the sustainability goals, he focused on how servitization can foster new technologies to help us become more sustainable and how innovation can help to cope with disruption. A smart equation was presented 'smart = resilient plus adaptive', and we can all contribute to 'smart' by considering systems and how they interact with each other, building in reliance through diversity and by taking the ecosystem view, decentralization of the

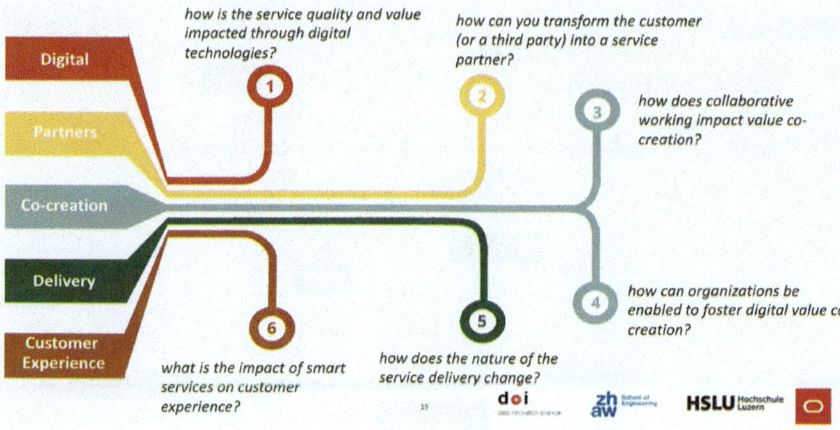

Fig. 4 Caption

system and creating systems that are self-repairing. Many of his views were described differently, yet in some way coherent with service-dominate logic.

Michael Blickenstorfer opened after lunch with 'smart services to support growing food'. He made a fascinating presentation on how we can grow food better using AI and, in doing so, contribute to eight of 17 of the UN's sustainable development goals. He is working on the technology and considering different servitization-based business models to support and accelerate the introduction of the technology into the market. The ongoing AI developments can be packaged into growing support services that reduce the unit inputs while maximizing the yield. The overview of the system is shown in Fig. 5.

Markus Wenig presented 'smart services to support shipping', describing how WIN GD had developed an engine health monitoring and management system for ships that focus on the engines they design. The business operates in a complex software landscape with many different systems, making it challenging for them to find space within the solutions offered. The inherent challenge that he described was communicating and assessing the system's value, which then impacts the customer's willingness-to-pay. Yet on the successful side, he confirmed that they have over 200 vessels connected and collect, on average, over 3GB of data per day. An essential challenge for shipping remains the connectivity due to the cost of data transfer over satellite which creates some 'interesting' challenges. He closed with the lessons learnt:

- **Why**. Value! Collaborate from the early concept phase customer-centricity
- **How**. Simple and effective solutions (and business ecosystems), MVPs that work. Combine capabilities from many different areas inside and outside the firm.

Prof. Dr. Christine Bratrich closed the summit with her keynote, 'a reminder of sustainability' in particular, she presented the 17 Sustainable Development Goals that were agreed by all of the UN member countries. Until COVID-19, there had

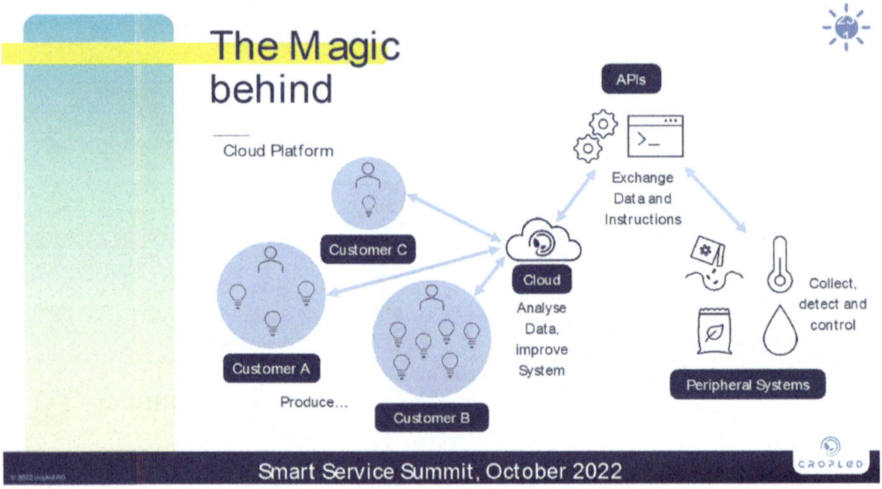

Fig. 5 The ecosystem that delivers the magic to optimize the yield

been good progress made in lifting people out of extreme poverty however, since then, the situation has become worse again. The impact today of global heating continues to create challenges for the wider sustainability goals, and coordinated action is needed to help us deliver the challenges we now face; therefore, the real question today is *how do we speed up the transformation?* Three cases describing actions that were improving sustainability were presented as inspiration: Synhelion—producing net-zero fuels; Planted—supporting plant-based diets, and oxara—closing the circle in the construction industry. Each of the cases draws upon the 17 Sustainable Development Goals as they provide a solid foundation for reporting sustainability as well as focusing on practical actions, in particular:

- **Be bold and connect**: We are facing several existential crises in parallel. Our solutions must bridge disciplines to tackle our challenges.
- **Align your ideas**: Despite their complexity, SDGs provide the strongest international framework. Let's use them: As strategic tools. As a communication tool. As reporting tool.
- **Accelerate the transformation**: Many of the younger generations are demonstrating how to push for rapid sustainability transformations. Join them and make a difference in your own professional lives!

The Summit in Graphics

Ramon Spaeti created a visual journey of the day to help us remember what we discussed and how it connected back to the summit's aim. The visualization of the designer helps (shown in Fig. 6) to simplify the message and create connections that otherwise practitioners and academics would miss.

Fig. 6 The summit in graphics

Reflections of the Editors

To provide an individual reflection on all of the individual papers would mean that readers might not read the papers, so as editors, we have all provided three reflections or learnings from the summit:

- Jürg—it's all about value creation: value for customers and providers and value for ecological goals needs to be considered in a combined way.
- Jürg—it is crucial to take into consideration social value for people, as technology cannot simply be introduced over people's minds and heads.
- Jürg—start with the purpose and goal in mind and let technology follow up these principles and goals.
- Shaun—we need to learn to integrate sustainability into the way we 'do business today, it is no longer an add on.
- Shaun—we need to think in (service) systems that are interconnected, and actions have consequences.
- Shaun—we need a diversity of perspectives and opinions to build smart services that really contribute to sustainability.

- Thierry—we need to assume that service users have zero time for and zero knowledge about our solution.
- Thierry—technology is rarely the issue. We need humans to agree on the meaning (semantics) and usage of services. Include deep domain experts.
- Thierry—make sure you include users and outsiders from time to time to ask inconvenient or 'naïve' questions.

Contents

About the Editors

Dr. Jürg Meierhofer is head of the "Smart Services" group of the data innovation alliance as well as director of studies MAS Industry 4.0 and CAS Smart Service Engineering at the ZHAW Zurich University of Applied Sciences. The optimization and design of data-driven value creation with smart services are a common thread throughout his activities. After holding various management positions in the service and innovation sector, he has been teaching and researching at ZHAW since 2014. He leads cooperation projects with numerous industrial companies and regularly writes publications. Jürg Meierhofer holds a Ph.D. from ETH Zurich and an Executive MBA from the University of Fribourg.

Prof. Dr. Shaun West after gaining a Ph.D. from Imperial College in London, Shaun worked for over 25 years in several businesses related to industrial services. He started his industrial career with AEA Technology before moving to National Power, where he developed and sold services to external businesses. After studying at HEC (Paris) for an MBA, he moved to GE Energy Services, modelling and negotiating long-term service agreements. At Sulzer, he drafted the strategy that led to the service division tripling in size over 10 years and executed part of the strategy by acquiring a 220M CHF service business. Now at the Lucerne University of Applied Sciences and Arts, he is the Professor of Product-Service System Innovation. He focuses his research on supporting industrial firms to develop and deliver new services and service-friendly business models. He is a member of the advisory board for ASAP Service Management Forum and a member of the data innovation alliance. He lives close to

Zurich with his wife and two children. He climbs, skis and runs. e-mail: shaun.west@hslu.ch

Dr. Thierry Buecheler is part of the 4-person management team running the Business Value & Strategy (management consulting) team across Oracle EMEA. He focuses primarily on the "North" and "East" regions, including Scandinavia, DACH, Central and Eastern Europe, and the Balkans. He builds on experience at the interface of business and IT in several sectors and industries. Before joining Oracle, he was an Engagement Manager at McKinsey & Company in Europe, North America and Africa and a visiting researcher at NEC Tokyo. Before that, he worked for two Swiss banks in the IT and strategy departments. He holds a Ph.D. in Computer Science/Artificial Intelligence from the University of Zurich and an M.Sc. in Computer Science from ETH Zurich. In his spare time, he spends as much time as possible with his family and outdoors, or he programs web and machine learning applications.

Empowering Sustainable Manufacturing Session

Value Perceptions on Smart Service Offerings in Manufacturing

Martin Ebel, Marleen Voss, Jens Poeppelbuss, Andreas Greve, André Sobieraj, and Frank Schomburg

Abstract Digital servitization describes the change of business models towards smart service offerings enabled by digitization. One challenge of digital servitization for manufacturers is to gain a better understanding how smart services generate value-in-use. To examine individual value-in-use concepts regarding smart services, we conducted and analyzed 22 in-depth interviews with the repertory grid technique. Our findings show that the respondents have a fairly uniform understanding of what constitutes an ideal service. Their perception of the service of tomorrow, smart service, and remote service offerings is closer to a future ideal than the service of today. Further we were able to group individual value perceptions into four groups and to discuss differences of an external and internal view on smart service. Interestingly sustainability was not mentioned as a value dimension in our study, which should be investigated further.

Keywords Smart service · Repertory grid · Value-in-use

M. Ebel (✉) · M. Voss · J. Poeppelbuss
Ruhr-Universität Bochum, 44801 Bochum, Germany
e-mail: Martin.Ebel@isse.ruhr-uni-bochum.de

M. Voss
e-mail: marleen.voss@isse.ruhr-uni-bochum.de

J. Poeppelbuss
e-mail: jens.poeppelbuss@isse.rub.de

A. Greve · A. Sobieraj · F. Schomburg
nextpractice-Institut für Komplexität und Wandel, 28217 Bremen, Germany
e-mail: a.greve@nextpractice.de

A. Sobieraj
e-mail: a.sobieraj@nextpractice.de

F. Schomburg
e-mail: f.schomburg@nextpractice.de

J. Meierhofer et al. (eds.), *Smart Services Summit*, Progress in IS,
https://doi.org/10.1007/978-3-031-36698-7_1

1 Introduction

Service infusion, also coined as servitization in the manufacturing sector, might describe a promising strategic pathway for manufactures that are recently focused on product-centric offerings (Baines et al., 2020; Brax et al., 2021; Gebauer et al., 2020). When the omnipresent smartification of products is considered here, there is also talk about digital servitization (Favoretto, 2022). Digital servitization describes a change of market offerings to smart service solutions enabled by digitization. Companies are getting engaged in service ecosystems and try to offer smart services instead of, or in addition to, their products. Such transformation concerns capabilities and processes and is taking into question organizational activities, especially in product-oriented industries like manufacturing (Kohtamäki et al., 2020). Besides a technology push, there seems to be an additional market pull for smart service offerings in the form of platform business models, outcome-based contracts and integrated solutions (Kohtamäki et al., 2019).

A challenge of digital servitization for manufacturers is to gain a better understanding of how smart service offerings actually generate value (Klein et al., 2018). Hereby the concept of value-in-use is challenging a prevailing view, that value is embedded in produced and exchanged manufactured goods (Grönroos & Ravald, 2011). Monitoring the value-in-use realized by the customer is mentioned as an important capability that suppliers need today, especially when it comes to co-development and co-creation of service offerings (Prohl & Kleinaltenkamp, 2020). Previous research has already studied the value-in-use of customer solutions that integrate products and services (Macdonald et al., 2016; Raja et al., 2013; Schaefers et al., 2021). However, an investigation of the value-in-use of smart service offerings in particular has not been carried out. Therefore, the aim of this study is to explore and understand the perception of values in smart service offerings. Here we would like to answer the question: *What are value perceptions on smart service offerings within the manufacturing industry?* For this purpose, a total of 22 interviews were conducted and analysed, based on the repertory grid technique.

In the next section we will give a short introduction of the used concepts and related works. In the method section we will introduce the interview technique. Following we will present our results and discuss them in relation to related works and possible further research directions.

2 Related Works

Service dominant (sd)-logic is diffusing into many different research fields by preparing a theoretical lens comprising a comprehensive view on value co-creation (Vargo & Lusch, 2017). In manufacturing this lens can help to describe the "incorporated paradigm shift from leadership in technology to leadership in use" (Meier et al., 2011, p. 1175) and to view smart products as boundary objects within value-co-creating service ecosystems (Beverungen et al., 2019).

Nonetheless, as Grönroos and Ravald (2011) note, the concept of value co-creation within sd-logic is too abstract for theoretical and practical analysis. Addressing the ongoing discussion of differently conceptualized understandings of value (Zeithaml et al., 2020) we follow a social constructivist approach where value is always co-created by the customer. Based on the value-in-use paradigm, value is understood as the effects perceived by a customer that result from resource integration and that facilitate or hinder goal achievement (Prohl & Kleinaltenkamp, 2020). The value for the customer is determined both by the benefits expected but also influenced by efforts and costs as well as required resources and capabilities when integrating resources of a supplier (Eggert et al., 2019). For purpose of innovation, service suppliers need to convince customers of a possible value contribution by communicating value propositions. We assume that by better understanding value perceptions, suppliers can offer superior value propositions with high potential value-in-use to their customers.

Prohl and Kleinaltenkamp (2020) emphasize that B2B companies need to perform activities to identify and record value-in-use (monitoring), but also perform activities to expand value-in-use and explore opportunities to increase customer value. Hereby individual value perceptions are a decisive factor for the acceptance of value propositions (Eggert et al., 2019), even noting, that smart service is a multi-actor phenomenon (Anke et al., 2020). For the analysis of individual values, for example, so-called hierarchical value chains can be investigated, which in their simplest form link attributes, consequences and personal values (Reynolds & Olson, 2001). The recent research by Schaefers et al. (2021) shows that the reputation of previous outcome-based contracts as an attribute is related to the avoidance of downtime (consequence) and can address the three superordinate goals (personal values) of reliability, organizational and individual security, and individual accountability (Schaefers et al., 2021). In the B2B context, Macdonald et al. (2016) distinguish between collective and individual values, which can represent the trade-off between values of an organization and individuals. In their study they find that organizations want to avoid downtime, remain competitive and reduce tied-up capital due to solutions business. Individuals on the other hand are valuing a reduction of pressure, reduced uncertainties and higher reputation (Macdonald et al., 2016).

Knowledge of such value elements can be used to gain a better understanding of customers' value-in-use and desired outcomes. For example, the consulting firm Bains & Company derives individual value elements from existing consulting projects. These include, for example, reduced effort, time savings, risk reduction, configurability, but also social responsibility as possible value elements in the B2B context. There are lists of such value elements for both the B2B (Almquist et al., 2018) and the B2C context (Almquist et al., 2016). Named examples show, that value elements in the context of servitization have been investigated throughout the last years. However, due to the progress of digitization and the emergence of smart services an investigation of the value-in-use of smart services has not been carried out.

3 Methodology

To examine individual value-in-use concepts regarding smart services within the manufacturing industry, we conducted 22 in-depth interviews with an interview technique that is based on the repertory grid (Kelly, 1955; Kruse et al., 2020; Tan & Hunter, 2002). The repertory grid is a technique for identifying the ways that a person construes (interprets or gives meaning to) his or her experience. It is capable to elicit difficult-to-articulate characteristics, i.e., the personal constructs of value (Macdonald et al., 2016).

Data gathering and analysis were supported by the nextexpertizer software (https://nextpractice.de/en/expertizer/). The interview technique nextexpertizer, developed by nextpractice, combines the advantages of quantitative and qualitative-empirical survey methods. It allows a qualitative-intuitive insight into the mostly unconscious evaluation structures and at the same time enables a mathematical comparability and condensation of the individual interviews as in questionnaires.

We conducted 13 interviews with employees of a German manufacturer of rotating equipment and nine with their customers. The company employs over 5000 people and supplies components and solutions for the building infrastructure. The interviews have a duration of 90–120 min. We asked the respondents about concepts related to smart service offerings. They described differences or similarities in their own words from their personal perspective. They intuitively assigned smart service elements to emerging rating scales, enabling individual value patterns to be identified (Kruse et al., 2020).

A study with nextexpertizer starts with the determination of comparison elements, which define the survey context. They form the associative framework for the survey and are intended to motivate the respondents to introduce topics relevant to them regarding the subject of the study. In total, 70 elements, considered relevant for the topic of smart service, were used for the survey. Exemplary items used for the interviews are: ideal service, ideal data handling, smart product, service in the past, service tomorrow, buying a product, leasing a product, performance-based contracting.

The conducted interviews proceed in three steps (see Fig. 1). At the beginning of a survey, two elements of the list are selected. The respondent is asked to classify the two elements as similar or different (*Compare*). Then, he/she describes the difference or similarity in his/her own words using a descriptive dimension that is personally meaningful to him/her (*Name*). In the third step, he/she then quickly and intuitively evaluates elements that represent the field of investigation (*Evaluate*).

Interviews were all done by professional interviewers. Figure 1 shows the software interface of nextexpertizer, used within the interviews. The entire procedure is repeated until the interviewee has introduced all the topics that are important to him/her. In total, each interviewee forms an average of seven rating scales and makes around 500 intuitive assessments in the interview.

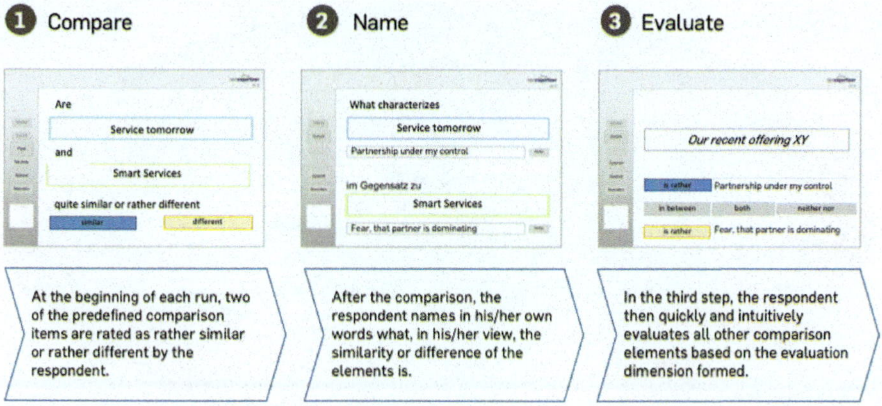

Fig. 1 Procedure of the interview

After conducting the interview with nextexpertizer, a matrix (Bertin display) is created, which forms the raw material of nextexpertizer's analysis. This matrix represents the essential aspects of the subjective perception of the individual respondent. In this rating matrix all formed rating scales and the assignments of the elements are recorded. These evaluation matrices are calculated via algorithms and generate a mathematically evaluable (quantitative) solution space. The individual rating matrices are mathematically comparable.

The condensation takes place purely on the level of the surveyed evaluation patterns, independent of the terms used by the interviewees. It provides an insight into the collective intuition and thus into the collective evaluation pattern. On this basis, groups can be formed, and types characterized. In contrast to other qualitative methods, the scope for interpretation is very small and the reliability (interrater reliability) of the aggregation steps is exceptionally high.

4 Results

The central data basis of this study is a calculated value space that represents the diversity of smart service experiences. The overall space emerges as a common evaluation space through the mathematical condensation of the individual interviews. In the diagram, a green area shows the position of the aspects rated positively (*desired properties*) by the interviewees, while a red area (*rejected properties*) contains the aspects rated negatively. Statements that are close to each other in the calculated value space are grouped into a cluster and described with an overarching term.

In this study, the 220 freely named original statements of the participants were condensed into 28 clusters. The semantic map is calculated based on different dimensions. On the east–west axis, on the one hand, there are person-related service aspects (west) and, on the other, technological aspects (east). The north–south axis, on the

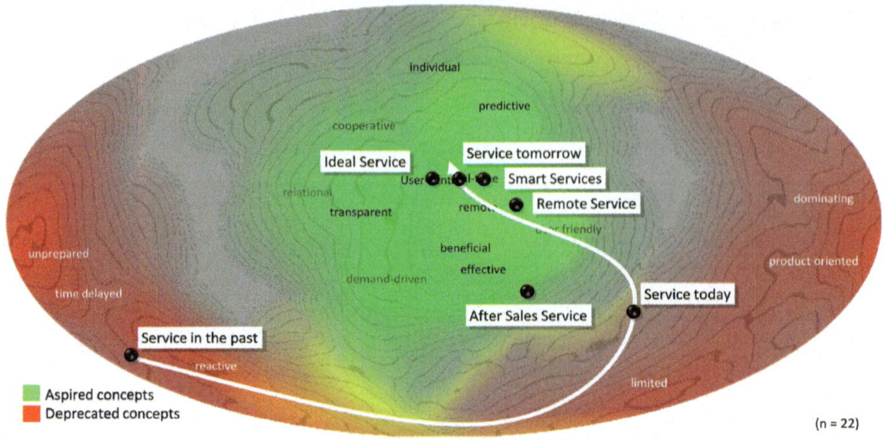

Fig. 2 Classification of past, current and future service

other hand, is about the role of the customer. Whereas the south is concerned with aspects demanded by the customer as the recipient of services, the topics in the north are concerned with the interaction between provider and customer.

The analysis of the data gives an insight into the differences of perceptions about service in the past, service today and service tomorrow. Overall, the respondents agree on what constitutes ideal service and that their perception of service tomorrow, as well as smart service and remote service, is closer to the ideal than service today (Fig. 2).

The evaluations show that the service of the past is strongly associated with personal on-site service. It is viewed critically because it was slow, reactive, externally determined and product-oriented and will therefore no longer meet the customer expectations of tomorrow. In particular, the customers expect manufacturers to provide professional solutions that take burdens from them and that they deal with potential problems in a spirit of partnership. They see value-in-use in predictive intelligent systems, which support them proactively and individually. Today's service is perceived as more effective, beneficial and forward-looking than service from the past. The customer relationship, on the other hand, is somewhat more formal than in the past. In general, a positive development towards transparency, real-time service and individualization of services is expected. The external respondents also expect services in the future that are less product-oriented and allow individual configurations.

As another finding, we were able to find first indicators of differences between internal perceptions (13 interviews with employees) and external perceptions (9 interviews with customers) of an ideal service. It shows that the internal respondents tend to be more concerned with the question of how customers can be involved in the development of new technologies in the future. The external respondents tended to be more concerned with the question of how services that are provided in personal contact can be automated.

Table 1 Identified groups and their value perceptions

Group	What is important to the group	What is rejected by the group
Co-development (n = 5)	• Develop together with customers • Direct remote access to data • Real-time access to systems	• Develop without customers • Maintenance-intensive approaches • Complicated procedure
Close relationship (n = 5)	• Partnership relationship • Offer a high value in use • User-friendly operation	• Formal customer relationship • React to faults with a time delay • Be inefficient
Customized solutions (n = 7)	• Simple and transparent processes • Understand the customers' needs • Relieve the burden with user-friendly solutions	• Communicate complicated and imprecise • Offer only standard technical solutions • Sell hardly to connect products
Maximum relief from burdens (n = 5)	• Be proactive in services • Be supported with intelligent systems • Enable individual configurations	• Reactive view of service • Ignore customers due to internal product focus • React passively to customer needs

Our findings also indicate that the respondents might be distinguished into four groups (Table 1), including those that emphasize the value generated by (1) the co-development of offerings, (2) a close relationship between provider and customer, (3) customized solutions, and (4) a maximum relief from burdens.

5 Discussion and Conclusion

By examining several employees as well as customers, it was possible to gain initial insights into the value dimensions of several stakeholders on the topic of smart service. Overall, our findings show that smart services are viewed positively by the respondents of our study. Smart service is closer to the ideal service and thereby it might be postulated, that the future of service lies in smart service. Customers expect manufacturers to provide professional solutions that take the pressure off them and to deal with problems in a spirit of partnership. Digital servitization thereby appeals to the values of suppliers and customers and, hence, is likely to remain a relevant research field.

Recently, the topic of digital servitization has been increasingly linked to the topic of sustainability. Sustainability is postulated as a driver and output of digital servitization (Abdelkafi et al., 2022; Parida & Wincent, 2019). Even though this is still a young field of research, it was surprising that sustainability did not play a role in the value perceptions about smart services of the interviewees. The connection between digital servitization and sustainability, which is postulated in science (Parida et al., 2019) and has been taking place for decades through the discussion of product service systems (Bocken et al., 2014; Kjaer et al., 2019; Tukker & Tischner, 2006), does not yet seem to have an impact on perceptions of ideal services in the future. It will be interesting to see whether this changes over the next few years.

Also, the respondents agree on what constitutes an ideal service. However, the internal and external respondents differ in their focus topics. While internally, emphasis is put on the process of customer involvement, customers look more closely at the outcome. Due to this finding, we provide an interesting distinction between provider and customer perspectives concerning the co-development of smart service offerings. While co-developing offerings are considered important by the suppliers, the customers expect the best possible solutions without being required to engage themselves too intensively. Investigating a more differentiated view of the customer's role seems therefore important for future research (Heinonen et al., 2010).

As our results are based on a small number of interviews only, further studies should be conducted to validate and extend our findings. It would be interesting to examine different industries to find out which similarities and perhaps also differences can be identified. It is possible that overarching value patterns may emerge across industries that can be used to develop new smart service offerings based on expected value dimensions of different customer groups.

References

Abdelkafi, N., Pero, M., Masi, A., & Capurso, I. (2022). Revisiting the servitization-sustainability link: A case study in the professional printing supply chain. *Cleaner Logistics and Supply Chain, 4*, 100061. https://doi.org/10.1016/j.clscn.2022.100061

Almquist, E., Cleghorn, J., & Sherer, L. (2018). The B2B elements of value. *Harvard Business Review, 11*.

Almquist, E., Senior, J., & Bloch, N. (2016). The elements of value—Measuring and delivering what consumers really want. *Harvard Business Review, 9*.

Anke, J., Poeppelbuss, J., & Alt, R. (2020). It takes more than two to tango: Identifying roles and patterns in multi-actor smart service innovation. *Schmalenbach Business Review, 72*(4), 599–634. https://doi.org/10.1007/s41464-020-00101-2

Baines, T., Ziaee Bigdeli, A., Sousa, R., & Schroeder, A. (2020). Framing the servitization transformation process: A model to understand and facilitate the servitization journey. *International Journal of Production Economics, 221*, 107463. https://doi.org/10.1016/j.ijpe.2019.07.036

Beverungen, D., Müller, O., Matzner, M., Mendling, J., & vom Brocke, J. (2019). Conceptualizing smart service systems. *Electronic Markets, 29*(1), 7–18. https://doi.org/10.1007/s12525-017-0270-5

Bocken, N. M. P., Short, S. W., Rana, P., & Evans, S. (2014). A literature and practice review to develop sustainable business model archetypes. *Journal of Cleaner Production, 65*, 42–56. https://doi.org/10.1016/j.jclepro.2013.11.039

Brax, S. A., Calabrese, A., Levialdi Ghiron, N., Tiburzi, L., & Grönroos, C. (2021). Explaining the servitization paradox: A configurational theory and a performance measurement framework. *International Journal of Operations & Production Management* (ahead-of-print). https://doi.org/10.1108/IJOPM-08-2020-0535

Eggert, A., Kleinaltenkamp, M., & Kashyap, V. (2019). Mapping value in business markets: An integrative framework. *Industrial Marketing Management, 79*, 13–20. https://doi.org/10.1016/j.indmarman.2019.03.004

Favoretto, C. (2022). From servitization to digital servitization: How digitalization transforms companies' transition towards services. *Industrial Marketing Management, 18.*

Gebauer, H., Fleisch, E., Lamprecht, C., & Wortmann, F. (2020). Growth paths for overcoming the digitalization paradox. *Business Horizons, 63*(3), 313–323. https://doi.org/10.1016/j.bushor.2020.01.005

Grönroos, C., & Ravald, A. (2011). Service as business logic: Implications for value creation and marketing. *Journal of Service Management, 22*(1), 5–22. https://doi.org/10.1108/09564231111106893

Heinonen, K., Strandvik, T., Mickelsson, K., Edvardsson, B., Sundström, E., & Andersson, P. (2010). A customer-dominant logic of service. *Journal of Service Management, 21*(4), 531–548. https://doi.org/10.1108/09564231011066088

Kelly, G. (1955). *Personal construct psychology*. Norton.

Kjaer, L. L., Pigosso, D. C. A., Niero, M., Bech, N. M., & McAloone, T. C. (2019). Product/service-systems for a circular economy: The route to decoupling economic growth from resource consumption? *Journal of Industrial Ecology, 23*(1), 22–35. https://doi.org/10.1111/jiec.12747

Klein, M. M., Biehl, S. S., & Friedli, T. (2018). Barriers to smart services for manufacturing companies—An exploratory study in the capital goods industry. *Journal of Business & Industrial Marketing, 33*(6), 846–856. https://doi.org/10.1108/JBIM-10-2015-0204

Kohtamäki, M., Parida, V., Oghazi, P., Gebauer, H., & Baines, T. (2019). Digital servitization business models in ecosystems: A theory of the firm. *Journal of Business Research, 104*, 380–392. https://doi.org/10.1016/j.jbusres.2019.06.027

Kohtamäki, M., Parida, V., Patel, P. C., & Gebauer, H. (2020). The relationship between digitalization and servitization: The role of servitization in capturing the financial potential of digitalization. *Technological Forecasting and Social Change, 151*, 119804.

Kruse, P., Greve, A., & Schomburg, F. (2020). *Next practice - erfolgreiches Management von Instabilität: Veränderung durch Vernetzung* (9. Auflage). GABAL.

Macdonald, E. K., Kleinaltenkamp, M., & Wilson, H. N. (2016). How business customers judge solutions: Solution quality and value in use. *Journal of Marketing, 80*(3), 96–120. https://doi.org/10.1509/jm.15.0109

Meier, H., Völker, O., & Funke, B. (2011). Industrial product-service systems (IPS2). *The International Journal of Advanced Manufacturing Technology, 52*(9–12), 1175–1191. https://doi.org/10.1007/s00170-010-2764-6

Parida, V., Sjödin, D., & Reim, W. (2019). Reviewing literature on digitalization, business model innovation, and sustainable industry: Past achievements and future promises. *Sustainability, 11*(2), 391. https://doi.org/10.3390/su11020391

Parida, V., & Wincent, J. (2019). Why and how to compete through sustainability: A review and outline of trends influencing firm and network-level transformation. *International Entrepreneurship and Management Journal, 15*(1), 1–19.

Prohl, K., & Kleinaltenkamp, M. (2020). Managing value in use in business markets. *Industrial Marketing Management, 91*, 563–580. https://doi.org/10.1016/j.indmarman.2020.03.017

Raja, J. Z., Bourne, D., Goffin, K., Çakkol, M., & Martinez, V. (2013). Achieving customer satisfaction through integrated products and services: An exploratory study: Achieving customer satisfaction through integrated products and services. *Journal of Product Innovation Management, 30*(6), 1128–1144. https://doi.org/10.1111/jpim.12050

Reynolds, T. J., & Olson, J. C. (Hrsg.). (2001). *Understanding consumer decision making: The means-end approach to marketing and advertising strategy*. L. Erlbaum.

Schaefers, T., Ruffer, S., & Böhm, E. (2021). Outcome-based contracting from the customers' perspective: A means-end chain analytical exploration. *Industrial Marketing Management, 93*, 466–481. https://doi.org/10.1016/j.indmarman.2020.06.002

Tan, F. B., & Hunter, M. G. (2002). The repertory grid technique: A method for the study of cognition in information systems. *MIS Quarterly, 26*(1), 39. https://doi.org/10.2307/4132340

Tukker, A., & Tischner, U. (2006). Product-services as a research field: Past, present and future. Reflections from a decade of research. *Journal of Cleaner Production, 14*(17), 1552–1556.

Vargo, S. L., & Lusch, R. F. (2017). Service-dominant logic 2025. *International Journal of Research in Marketing, 34*(1), 46–67.

Zeithaml, V. A., Verleye, K., Hatak, I., Koller, M., & Zauner, A. (2020). Three decades of customer value research: Paradigmatic roots and future research avenues. *Journal of Service Research, 23*(4), 409–432. https://doi.org/10.1177/1094670520948134

Communicating the Value of Digital Transformation Within Manufacturing Firms

Matthew Anderson, Shaun West, David Romero, and David Harrison

Abstract Digital Transformation provides both value creation and co-creation opportunities within manufacturing firms, enabling these to develop new and more efficient 'digitally-enabled' value propositions. However, there remain notable gaps in communicating these digital opportunities (value) throughout a firm, creating barriers to justifying (digital) transformations, and limiting their adoption. This paper aims to explore this gap using input from industrial case studies and to create a framework that recognises relevant metrics to identify and communicate digital value. The findings highlight the importance of translating metrics between the different levels of a firm and communicating visually these to better appreciate the new digitally-enabled value (co-)creation process(es).

Keywords Value creation · Digital transformation · Service-dominant logic

1 Introduction

Digital Transformation is a phenomenon where firms use digital technologies to modify or improve their existing business process to innovate their value creation, delivery, and/or capture mechanisms (Hess et al., 2016). *Digitalization* can create significant value, but such innovations come with risks (Anderson et al., 2021).

M. Anderson (✉) · D. Harrison
Glasgow Caledonian University, Glasgow, Scotland
e-mail: mander237@caledonian.ac.uk

D. Harrison
e-mail: d.f.harrison@gcu.ac.uk

S. West
Lucerne University of Applied Sciences and Arts, 6048 Horw, Switzerland
e-mail: shaun.west@hslu.ch

D. Romero
Tecnológico de Monterrey, Mexico City, Mexico
e-mail: david.romero.diaz@gmail.com

J. Meierhofer et al. (eds.), *Smart Services Summit*, Progress in IS,
https://doi.org/10.1007/978-3-031-36698-7_2

Business cases have become the best practice when developing digital solutions for new or existing processes (Colli et al., 2021). User stories may be created to support them and their adoption (Shaughnessy, 2018). This is distant from the lean transformation approach, where every process improvement project requires an A3 sheet (Schwagerman & Ulmer, 2013). The lean philosophy focuses on 'value' and 'waste' and requires performance metrics to help assess the value. Service-Dominant Logic (SD Logic) can provide insights into value creation and co-creation (Grönroos, 2011; West et al., 2018).

This paper will describe three digital solutions and their enabling Industry 4.0 technologies within different production environments, using a lean thinking approach and SD logic to explain their value (co-)creation process. The aim is to establish a new framework to support *value descriptions* that may be considered as the perceived benefits delivered to the customer.

2 Literature Review

Shook (2009) described how *performance management* could be achieved sustainably by applying the *"Toyota A3" (lean) method/tool.* Bellisario and Pavlov (2018) examined the use of *lean principles* for manufacturing firms and confirmed the value of 'lean' for *continuous improvement.* The link between *operational performance management* and *knowledge building* (via *action-learning*) has been described by Staedele et al. (2019). *Business performance* can be measured through performance metrics (Neely et al., 1995), and these metrics are proxies for the value-creation processes of the firm. *Metrics* need to be relevant to the different organizational levels and stakeholders (e.g., production line and plant) within the firm (Huang & Badurdeen, 2018); otherwise, they become detached from the value creation processes. The *metrics* should consider *input*, *process*, and *outcome indicators*, which can be measured to allow the operational levers that support the value creation process to be evaluated.

Anderson and Narus (1998) stated that it is important to understand "what customers value in a B2B environment?". This can be estimated from *metrics* (Anderson, 1995), and it is further necessary to understand the individual firm's operation to consider the value creation offered by a *value proposition.* A *value creation process* can be defined from a service science perspective (Grönroos, 2017), where its details should be broken down (into activities) to be fully understood; this is in line with the approach proposed by Anderson and Narus (1998). Within this paradigm, "value (co-)creation" and "value destruction" within a firm should be understood and their value estimated (Bruce et al., 2019). This is relevant to an Industry 4.0 production environment (Bordeleau et al., 2019), where detailed insights into value (co-)creation/destruction can be understood and quantified.

Service-Dominant Logic (SD Logic) is derived from the service science community (Grönroos, 2017; Vargo & Lusch, 2018). The link for this paper is the concept of "value-in-use", and around this, we can describe the activities, actors, and institutional arrangements that support a value (co-)creation and destruction process(es). The use of *SD logic* to support an (Industry 4.0) technology-driven service innovation has been described in the scientific literature (Blaschke et al., 2019; West et al., 2018). The approaches have organized the foundational premises of *SD logic* into a form that can support the design of systems that support value co-creation, in effect through different forms of resource integration made possible by the institutional arrangements that *digital systems* provide. This switch to *SD logic* has been considered disruptive (Chester Goduscheit & Faullant, 2018), as it provides a change from the technology approach where value is confused with functionality (Anderson & Narus, 1998). An approach to describe the ontology of the 'multiple' value co-creation occurrences has been developed by Blaschke et al. (2018).

3 Research Methodology

The approach of multiple short case studies with further cross-case analysis has been chosen as the research methodology for this study (Bell et al., 2022). In this paper, three short cases, all from production environments, have been chosen. To reduce the risk of research bias, all the authors of this paper challenged the cases. The cases are introduced in the results section of this paper for context, they are then categorized for value creation or co-creation according to the dimensions in Table 1, which have been identified from the literature. The dimensions chosen originated from the lean, performance management, and SD logic body of knowledge to categorize the case studies' insights (Voss et al., 2002).

Table 1 Basic data capture for the cases

Category	Dimension	Description
General	Contextual description	• Problem description • Production purpose
Lean	Value creation	• What is the value that accrues?
Lean	Performance metrics	• What metrics can be used as proxies for value creation?
Marketing	Actors	• Identification of actors and roles
Marketing	Beneficiary(ies)	• Who benefits?
Marketing	Situation(s)	• When and where does value accrue?
SD logic	Value co-creation	• Value creation between actors • Interactions between diverse actors • Accommodation of roles • Resource integration
SD logic	Service platform	• Rules of exchange

4 Results

Case 1 "*Optimization of assembly activities in configured-to-order production*" is from an industrial equipment manufacturer that assembles pre-manufactured configured-to-order equipment for the process industry (see Table 2). The firm wanted to improve its existing process of *production order processing* in the factory. They tended to improve at the individual workstation level rather than at the production system level, which is against the balancing and levelling principles of lean. Their current system focuses on resource planning, with multiple spreadsheets and paper-based side systems covering manufacturing. This approach delays operators in getting the correct information at the right time to perform their work.

The approach taken by the firm was to implement a Manufacturing Execution System (MES) to provide information directly to the operators. The data flows could then be used for ongoing adaptations to their production planning process, improving its reliability. This led to better process discipline and orchestration of factory floor activities. It enabled a more targeted capital investment with data-driven (intelligent) maintenance and asset replenishment activities. These digitally-enabled, data-driven improvements proved challenging to quantify, as before implementation, it was not possible to get a baseline across the multiple operations management systems.

Case 2 "*Operational best insights from cloud-based smart factory*" considers a firm that provides process engineering solutions via make-to-order production (see Table 3). The firm experienced relevant issues with limited visibility of its global

Table 2 Case 1—optimization of assembly activities in configured-to-order production

Dimension	Description
Value creation	• Configured-to-order process: goods received, machining, assembly, tests, and delivery
Performance/waste metrics	• External: on-time delivery (waiting), and product performance • Internal: lead times, stoppage recording and analysis, identification of constraint points (motion and waiting), cost of rework (defects and over-processing)
Actors	• Installer procurement, project manager, manufacturing manager, factory floor supervisor/scheduler, operator, and suppliers
Beneficiary(ies)	• Installer, sales, equipment operator, equipment owner, and project manager
Situation(s)	• Beginning-of-Life (BOL) and early Middle-of-Life (MOL)
Value co-creation	• Manufacturing manager/production scheduler: ERP planning; optimized routings; and improved resource usage • Finance manager/production manager: optimized capital investment • Project manager/installer: improved coordinated installation • Manufacturing manager/supplier: order forecasting and inventory optimization
Service platform	• Standard protocols for manufacturing execution defined (and refined) • MES platform for sharing information on the status of different actors

operations. Coordination efforts are limited to monthly reports, assembled manually. This was found as too infrequent to allow for root-cause analysis and corrective measures to be actioned. Machine data was generated by the production machines but this had only been used for control, not monitoring capabilities.

The firm chose to use its machines' data captured by Programmable Logic Controller (PLC) systems on its production equipment/lines, aggregating it via an edge device and automatically transferring it to a cloud environment. The cloud provided the needed monitoring capability and alerts were targeted at different levels of the business, from operators to supervisors to shop floor managers. The cloud-based system offered digital Andon alerts, throughput information, and production bottleneck data. It also provided global visibility, showing cross-site production line comparisons, extended to the whole enterprise, with comparisons across product lines, actual costs of production, and site load balancing. It achieved this by combining operational and Enterprise Resource Planning (ERP) data to identify waste and provide insights for continuous improvement towards best practices.

Case 3 "*Overall Equipment Effectiveness (OEE) for legacy machine tools*" describes a manufacturer of medium volume configured-to-order industrial equipment that has multiple facilities with large machine parks of legacy machine tools (see Table 4). The majority of the legacy machines could not generate operational data, and there was a lack of infrastructure to collect, transfer, and process any data produced. Replacing or upgrading the legacy machines was not considered cost-effective, nor was installing a modern PLC system. The firm logged manually that they had problems with machine availability, performance, quality, yield and throughput.

The approach was to develop a solution in-house based on standard IoT sensors and PC components, available off-the-shelf at low cost as a proof-of-concept.

Table 3 Case 2—operational best insights from cloud-based smart factory

Dimension	Description
Value creation	• Local and global operations optimization
Performance/waste metrics	• Internal local operations: line OEE; back-logs size; and productivity • Internal global operations: site OEE; back-logs size; and productivity comparison (i.e., overproduction, over-processing, defects, and waiting)
Actors	• Global finance director, global operations director, local operations manager, and operators
Beneficiary(ies)	• Global finance, and local production management
Situation(s)	• Beginning-of-Life (BOL)
Value co-creation	• Global finance/head global operations: production costs optimization • Global operations/local operations manager: best practice sharing; and productivity improvement • Local operations manager/operators: reduction of defects; focus on more critical tasks; and reduction of other wastes
Service platform	• Global and local operations can share performance knowledge on a common platform

Table 4 Case 3—Overall Equipment Effectiveness (OEE) for legacy machine tools

Dimension	Description
Value creation	• Configure-to-order production, machining and assembly process
Performance/waste metrics	• Manufacturing performance: OEE breakdown, lead time, and quality (in terms of over-production, over-processing, defects, and waiting wastes)
Actors	• Factory supervisor, planner, operator, and local director
Beneficiary(ies)	• Factory supervisor, planner, operator, and local director
Situation(s)	• Beginning-of-Life (BOL)
Value co-creation	• Factory supervisor/operator: targeted production support, and improved productivity • Factory supervisor/planner: more accurate production capacity and scheduling • Factory supervisor/local director: improved on-time delivery, and better investment focus
Service platform	• Exchange protocols are not yet established; the solution needs to be embedded within an established continuous improvement methodology

These 'boxes' were attached to the legacy machines to generate data, alongside a Human–Machine Interface (HMI). The device was extended with additional interfaces capable of communicating directly with data-generation-capable machines. An Andon light tower was added, and it was IoT-enabled to generate push alerts to specific operators, providing transparency for availability, and with pre-populated stoppage-reason codes added from the operators' HMI for their selection.

5 Discussion

The dimensions provided a framework that gave insights into the case studies and helped describe the value created and its value co-creation process. Lessons from each dimension are described in Table 5. The framework supported an understanding of the value co-creation process and the context in which it occurred. The 'case-actor matrix' (Stoll et al., 2020), coupled with sketches of the interactions, would have further supported this description. The translation of the metric from meso to micro also needs additional support. The internal/external value creation model is helpful with value descriptions and linking them to the firm's KPIs.

The *Smart Factory vision* provides a challenging environment for describing value creation on an individual case basis. This is due to the potential for multiple beneficiaries within the system and multiple actors interacting in different situations during production operations management. Due to this, the *value co-creation processes* are complex to describe. In contrast, at the firm level may be significantly higher and easier to describe. The value creation may accrue as intangibles that support the firm's sustainability yet initially create minimal tangible value.

Table 5 Lessons from each dimension based on insights from the cases

Dimension	Description
Value creation	• This described the 'purpose' of the process within the business context
Performance/waste metrics	• Described how the business measured 'value'. It needed to be translated into terms that made sense at the individual actor level, confirming a need to describe the value at both the micro- and meso-levels. Here the use of lean was supportive at the micro-level
Actors	• It was important to be very specific here and to move beyond the simple terms 'customer' or 'user' to clearly define each actor and their role
Beneficiary(ies)	• It was necessary to align the value with the beneficiary, and the framework did not support this, rather the framework supported the identification of the beneficiary. In some cases, the beneficiary was not directly involved in the value co-creation process. This is not in alignment with SD logic
Situation(s)	• This helped with the descriptions, but the three simple options provide limited insights, so further details are needed
Value co-creation	• This provided short descriptions of the value creation process, nevertheless, simple dyadic/triatic sketches might help better describe the process; they could also support the storytelling of the actual process, and its actors' identification
Service platform	• This aspect provides institutional arrangements (needed for SD logic) and also provides insights into what to improve

The key lesson is to clearly define the problem, the waste, the actors, and the beneficiaries within a smart factory. An approach that can support the problem description is visualization, which supports decision-making by making the intangibles visible at the individual and firm levels. This aligns with Toyota's findings, where their approach creates a "learning organization" adaptable to change.

A clear problem description that describes both the waste (value destruction) and the value (co-)creation supports the development of key performance metrics (both input process, and outcome). The use of *SD logic* and *lean thinking* is supportive in a factory environment, yet a clear framework is needed to support the description and help managers understand the value of a co-creation process (Abid et al., 2022; Anderson, 1995). Toyota used its *A3 sheets* (Liker, 2004) to support the learning and knowledge-sharing processes. The approach also benefited from problem visualization, yet it failed to capture the actors and beneficiaries necessary to describe the benefits of value co-creation processes (West et al., 2021).

There is a clear need to communicate value throughout a firm (Rösler & Friedli, 2021). There are three parts to this: (i) the problem, (ii) the solution, and (iii) the impact. All parts need to be communicated, different languages need to be considered and a shared understanding must be gained. This paper provides insights into how this may be achieved. Theoretically, we must develop and test approaches to describe value accrual in a network. The approach must combine the *operational language* with *management metrics* and the *value co-creation concepts* from SD logic. A visual framework needs to be developed and tested, along with considerations of *sustainability*. Social and environmental impacts are more widely recognised as important dimensions of value, and are a key consideration for customers, partly to support their environmental responsibilities but also to encourage environmental accountability throughout a value co-creation platform (Polat, 2022). This direction may provide new value communication opportunities.

6 Conclusions and Recommendations

The aim to establish a framework to support metric selection/building has been applied in three cases. A link between actions that support value co-creation at the individual level and the firm's value accrual was needed. Understanding the value co-creation process (i.e., how, who, when, where, and what) is not trivial. For this reason, it is recommended that a visual is used to support the understanding of the value co-creation process and to identify the beneficiaries in each instance. We have seen a need for a 'Rosetta stone' to help translate the individual metrics to metrics at the firm's level to improve their relevance. It is recommended to develop a visual framework and test it with the three cases. This then could be further refined through development/testing loops.

References

Abid, M. F., et al. (2022). Value creation or value destruction: Conceptualizing the experiential nature of value-in-use. *Consumer Behaviour, 21*(3), 583–601.

Anderson, J. C. (1995). Relationships in business markets: Exchange episodes, value creation, and their empirical assessment. *Academy of Marketing Science, 23*(4), 346–350.

Anderson, J. C., & Narus, J. A. (1998). Business marketing: Understand what customers value. *Harvard Business Review, 76*, 53–67.

Anderson, M., West, S., & Harrison, D. (2021). Exploring accidental digital servitization in an industrial context. In *Advances in production management systems* (pp. 126–135). Springer.

Bell, E., Bryman, A., & Harley, B. (2022). *Business research methods*. Oxford University Press.

Bellisario, A., & Pavlov, A. (2018). Performance management practices in lean manufacturing organizations: A systematic review. *Production Planning & Control, 29*(5), 367–385.

Blaschke, M., Haki, K., Aier, S., & Winter, R. (2018). Value co-creation ontology—A service-dominant logic perspective. *Multikonferenz Wirtschaftsinformatik, 2018*, 398–409.

Blaschke, M., et al. (2019). Design principles for digital value co-creation networks: A service-dominant logic perspective. *Electronic Markets, 29*(3), 443–472.

Bordeleau, F. E., Mosconi, E., & de Santa-Eulalia, L. A. (2019). Business intelligence and analytics value creation in Industry 4.0: A multiple case study in manufacturing medium enterprises. *Production Planning & Control, 31*(2–3), 173–185.

Bruce, H. L., Wilson, H. N., Macdonald, E. K., & Clarke, B. (2019). Resource integration, value creation and value destruction in collective consumption contexts. *Business Research, 103*, 173–185.

Chester Goduscheit, R., & Faullant, R. (2018). Paths toward radical service innovation in manufacturing companies—A service-dominant logic perspective. *Product Innovation Management, 35*(5), 701–719.

Colli, M., Stingl, V., & Waehrens, B. V. (2021). Making or breaking the business case of digital transformation initiatives: The key role of learnings. *Manufacturing Technology Management, 33*(1), 41–60.

Grönroos, C. (2011). Value co-creation in service logic: A critical analysis. *Marketing Theory, 11*(3), 279–301.

Grönroos, C. (2017). On value and value creation in service: A management perspective. *Creating Value, 3*(2), 125–141.

Hess, T., et al. (2016). Options for formulating a digital transformation strategy. *MIS Quarterly Executive, 15*(2).

Huang, A., & Badurdeen, F. (2018). Metrics-based approach to evaluate sustainable manufacturing performance at the production line and plant levels. *Cleaner Production, 192*, 462–476.

Liker, J. K. (2004). *The Toyota Way: 14 management principles from the world's greatest manufacturer*. McGraw-Hill.

Neely, A., Gregory, M., & Platts, K. (1995). Performance measurement system design. *Operations & Production Management, 15*(4), 80–116.

Polat, I. (2022). Environmental sustainability, value co-creation, and innovation in service industries with the lens of S-D logic. In *Social customer relationship management (Social-CRM) in the era of Web 4.0* (pp. 128–152).

Rösler, J., & Friedli, T. (2021). A capability model for equipment-as-a-service adoption in manufacturing companies. *Progress in Information Systems*, 59–71.

Schwagerman, W. C., & Ulmer, J. M. (2013). The A3 lean management and leadership thought process. *Technology, Management, and Applied Engineering, 29*(4), 1–10.

Shaughnessy, H. (2018). Creating digital transformation: Strategies and steps. *Strategy & Leadership, 46*(2), 19–25.

Shook, J. (2009). Toyota's secret. *MIT Sloan Management Review, 50*(4), 30.

Staedele, A. E., Ensslin, S. R., & Forcellini, F. A. (2019). Knowledge building about performance evaluation in lean production. *Manufacturing Technology Management, 30*(5), 798–820.

Stoll, O., West, S., et al. (2020). Upgrading the Data2Action framework: Results deriving from its application in the printing industry. In *Exploring service science* (pp. 273–286).

Vargo, S., & Lusch, R. (2018). *The SAGE handbook of service-dominant logic*. Published.

Voss, C., Tsikriktsis, N., & Frohlich, M. (2002). Case research in operations management. *Operations & Production Management, 22*(2), 195–219.

West, S., Gaiardelli, P., & Rapaccini, M. (2018). Exploring technology-driven service innovation in manufacturing firms through the lens of service-dominant logic. *IFAC-PapersOnLine, 51*(11), 1317–1322.

West, S., et al. (2021). Digital twin providing new opportunities for value co-creation through supporting decision-making. *Applied Sciences, 11*(9), 3750.

Sustainable Value Optimization by Smart Services Along the Customer Lifecycle

Jürg Meierhofer and Melissa Stucki

Abstract This paper investigates the creation of economic and ecological value in manufacturing ecosystems. The focus is put on the B2B relationship during a lifecycle between a provider of production machines and a customer applying this machine in its own production processes. Smart services for manufacturing environments have been proven to create economic value for the actors in an ecosystem and there are several quantitative approaches to assess this value. Additionally, smart services have the potential to create ecological value by manifold levers such as, for example, higher equipment efficiency, extended lifetime, or more efficient maintenance processes. This paper extends the existing quantitative models for economic value creation by incorporating a quantitative model for ecological benefits and costs.

Keywords Smart services · Customer lifecycle · Value creation · Sustainability

1 Introduction

For manufacturers, the service business has the potential to create additional revenue and higher customer loyalty by extending their product business and by more stable cash flows (Ebeling et al., 2014). The customer gets additional benefits having an output from the equipment that is better targeted at its jobs to be done (Kowalkowski & Ulaga 2017). Economic value creation by smart services in manufacturing ecosystems is extensively documented in the research literature, e.g., Rapaccini and Adrodegari (2022), Schüritz et al. (2019), Zheng et al. (2020).

According to Tukker (2015), product service systems (PSS) represent an integrated bundle of products and services which aims at creating customer utility and generating value. They help a business to establish a more strategic position in the value network and enable it to capture more value. This can lead to higher cus-

J. Meierhofer (✉) · M. Stucki
Zurich University of Applied Sciences, CH 8401, Winterthur, Switzerland
e-mail: juerg.meierhofer@zhaw.ch

M. Stucki
e-mail: melissa.stucki@zhaw.ch

J. Meierhofer et al. (eds.), *Smart Services Summit*, Progress in IS,
https://doi.org/10.1007/978-3-031-36698-7_3

tomer loyalty and a deeper understanding of customers. Consequently, there is also a greater potential for service innovation. Additionally, PSS have the potential to create environmental value, e.g., by improved maintenance leading to better energy and resource utilization. The environmental effect is best achieved with output-oriented service models, because the use of materials and energy become a cost factor that the provider wants to minimize. PSS are considered among the most suitable business models for a circular economy (CE) (Kühl et al., 2018). Dey et al. (2022) investigate how CE practices positively impact the triple bottom line (economic, environmental, and social performance). Approaches for increasing material efficiency (such as, e.g., reduce, reuse, recycle) are discussed in Tecchio et al. (2017). Servitization approaches of product business helps companies to embrace their responsibility in the area of social and environmental sustainability by thinking more in terms of upgradable long-term solutions (Vendrell-Herrero et al., 2021).

Jasiulewicz Kaczmarek and Gol (2019) discuss in detail how smart services such as remote service or maintenance can improve the triple bottom line. Special attention is paid to the information flow along the lifecycle of the product. Downtimes of machines can be reduced, energy efficiency can be increased, and the operational lifetime can be extended substantially. According to Bertoni (2019), business value and sustainability value are often treated as two opposite objectives. To counteract this, they suggest to use multi-criteria decision making for the design of sustainable product-service systems and introduce a process scheme based on qualitative assessment of customer value, provider value, and the ecological impact.

There is a need for quantitative measures that combine the economic and ecological domain (Doni et al., 2019). The ecological benefit of smart services is conceptually and qualitatively analyzed and described in the literature, e.g., Bressanelli et al. (2018), Tecchio et al. (2017). There are models for the assessment and quantification of the economic value of data-driven services (Meierhofer, 2021; Meierhofer et al., 2022). Bertoni (2019) provides a model for the combined investigation of economic and ecological value on a qualitative level. However, in order to manage the environmental impact, manufacturers must align their servitization reportings better with their environmental, social, and governance (ESG) goals. Hence, there is a research gap in the field of quantitative models for the combined economic and ecological value creation in manufacturing service ecosystems. The goal of this paper is to reduce this gap. The research question of this study is: "*Which combination of services along the equipment lifecycle optimizes the economic and the ecological value in a combined way?*". A conceptual model is developed for answering this research question and a hypothetical numerical example is constructed for its application.

2 Methodology

This research builds on and extends the existing quantitative modelling approaches for economic value creation by smart services that are described in the literature. Additionally, the theoretical foundations for the ecological impact of smart services as well as for the quantitative indicators of their ecological impacts are described based on literature. Given this, the two domains are combined to an integrated quantitative model for both economic and ecological value creation. The model is evaluated numerically by an example case study. Service settings for the optimization of the value creation are discussed. We define the system boundaries in the incremental impact of smart services in the operational lifecycle of an equipment, i.e., we do not consider the effort for developing the equipment and services and the basic operation of the equipment without the additional services.

3 Quantitative Model

3.1 Economic Value Model

Building on Meierhofer et al. (2022), a model is used which considers the value created for the customer and the value captured for the provider as two independent dimensions which are jointly optimized by multi-objective optimization. In agreement with Bartels and Jenkin (1977) the current study further develops this model by integrating the societal goals as a third dimension, thus creating the basis for the integration of the ecological value.

The approach used in Meierhofer et al. (2022) quantitatively models the value creation along four lifecycle phases as shown in Fig. 1, where the phase "Initiate" represents the pre-sales, "Expand" the phase in which the customer gets acquainted with the new equipment, "Stabilize" the potentially many years long phase in which the equipment is operated and creates value but also needs maintenance, and "Terminate" the phase in which the customer wants to replace or give up the equipment and where services for reusing material or components or extending the lifetime come into place. The economic value for the provider (value capture) and the customer (value creation) is calculated by

$$V_P = \sum_{i=1}^{4} V_{P,i} \quad \text{and} \quad V_C = \frac{1}{CLT} \sum_{i=1}^{4} V_{C,i} \tag{1}$$

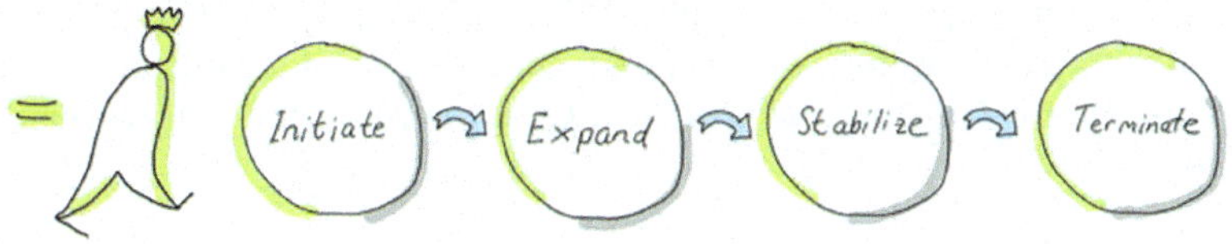

Fig. 1 Phases of the customer lifecycle based on Meierhofer et al. (2022)

where V_P is the value for the provider and each $V_{P,i}$ represents the value accrued in phase i of the lifecycle. Analogously, V_C and $V_{C,i}$ for the customer, whereby V_C is normalized by the customer lifetime for reasons explained in Meierhofer et al. (2022). The values are relative to the base scenario with no additional services activated (for service intensity 1 in the second column in Table 1) and therefore can become negative when the additional service creates more costs than benefit (e.g., in Fig. 2i).

3.2 Ecological Value Model

This paper extends this approach by incorporating the ecological value impact of the smart services along the lifecycle. Comparable to Achterberg et al. (2016), ecology is considered an additional stakeholder or actor in the service ecosystem. For the sake of simplicity of the model, the ecological impact is aggregated to one dimension by adopting the key parameter "CO_2 equivalent", CO_2e (Brander and Davis, 2012).

$$V_{Eco} = \frac{1}{CLT} \sum V_{Eco,i} \tag{2}$$

Analogously to Eq. 1, the impact is represented as an ecological value and is calculated over the customer lifecycle by Eq. 2 with $V_{Eco,i}$ being the ecological value contribution in terms of CO_2e reduction in lifecycle phase i and CLT the customer lifetime. Smart services can contribute to reducing CO_2e, e.g., by avoiding unnecessary logistics, thus increasing V_{Eco}. On the other hand, smart services introduce additional ecological impact, e.g., by the operations of sensors, actors, connectivity, or data processing, thus increasing CO_2e and reducing V_{Eco}.

Table 1 describes the modelling approach how smart services impact the ecological value in the four phases of the lifecycle by extending the concept of the economic value introduced in Meierhofer et al. (2022). The description of the ecological value is based on Bressanelli et al. (2018). The quantification of the $V_{Eco,i}$ components takes into account the reduction of the CO_2e impact by the services. Variables $\alpha_{1,2,3}$ denote the travel reduction by the smart services in phases 1, 2, and 3. $CO_2e_{travel,scrap,spare,machine}$ are the factors for calculating the CO_2 equivalent of these impacts. The amount of scrap parts or material saved in phases 2 or 3 is described by $\Delta scrap_{2,3}$. The lifetime extension of spare parts which need be replaced regularly several times during operations (i.e., phase "Stabilize") is described by ΔT_{sp} and the extension of the entire lifetime of the equipment at the end of the lifecycle by ΔCLT with α_4 used for calculating the positive CO_2e impact of the longer usage of the machine.

Table 1 Economic and ecological value creation in the four phases of the lifecycle. Note: in all phases, there are potentially additional economic value components given by service delivery costs (financial and environmental) and service revenues, which are not listed separately in the table due to space limitations. The negative ecological impact imposed by the services (additional energy and raw material like sensors etc. consumption) is taken into account by the variables $\Delta\ CO_2e_{IoT_{1,2,3,4}}$

Lifecyle phase	Service intensities for economic value $V_{P,i}$, $V_{C,i}$	Additional Impact on Ecological Value $V_{Eco,i}$
Initiate	2. Target offering based on data about the customer's need increases the conversion rate of sales 1. Not target offerings	Travel to the customer and other marketing logistics can be reduced thanks to higher conversion rate $V_{Eco,1} = \alpha_1 \cdot CO_2e_{travel} - \Delta CO_2e_{IoT_1}$
Expand	2. Higher performance thanks to targeted training for the customer based on data 1. No targeted training	Less material loss and scrap parts thanks to targeted training.Potentially more travel to the customer for targeted training. $V_{Eco,2} = \alpha_2 \cdot CO_2e_{travel} + \Delta scrap_2 \cdot CO_2e_{scrap} - \Delta CO_2e_{IoT_2}$
Stabilize	1...5: Improving the performance for the customer with 5 different service intensities (5. performance optimization 4. condition based maintenance 3. remote service 2. monitoring 1. standard)	Less material loss and scrap parts thanks to different levels of maintenance and optimization services Travel to the customer reduced thanks to remote services $V_{Eco,3} = \alpha_3 \cdot CO_2e_{travel} + \Delta scrap_3 \cdot CO_2e_{scrap} + \Delta T_{sp} \cdot CO_2e_{spareParts} - \Delta CO_2e_{IoT_3}$
Termi-nate	2. Targeted retention measures based on data insights from the customer increase the customer lifetime CLT and thus the customer lifetime value for the provider 1. No targeted retention	The extension of CLT by targeted offerings (ΔCLT) includes measures such as refurbishing or equipment right sizing, thus increasing the longevity of material and production output per material $V_{Eco,4} = \alpha_4 \cdot \Delta CLT \cdot CO_2e_{machine} - \Delta CO_2e_{IoT_4}$

4 Model Application

The application of the concept described in Sect. 3 is implemented in a numerical model. The economic value creation per lifecycle phase is reflected by applying the schemes and the numerical example described in Meierhofer et al. (2022) and enhanced by the ecological component to a three-dimensional optimization problem as shown in Fig. 3. Typical values for CO_2e are obtained from Ecoinvent (2022). Referring to the research question formulated in this study, the numerical model evaluates all 40 ($= 2 \cdot 2 \cdot 5 \cdot 2$) different combinations of services along the lifecycle

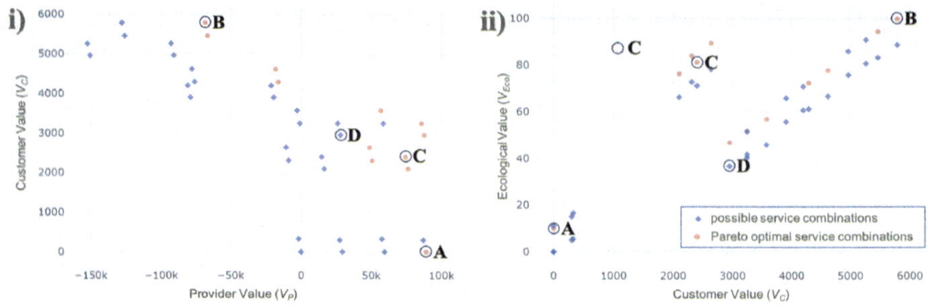

Fig. 2 Two-dimensional projections of the scatter plot of Fig. 3

as described in Table 1. The resulting values V_P, V_C, and V_{Eco} are shown in the three-dimensional scatter plot in Fig. 3.

As discussed in Meierhofer et al. (2022), the optimum service configurations can be identified on the Pareto front of the values for the customer, for the provider, and for ecology for the different service combinations. On the Pareto front, improving one value component (e.g., V_C) can only be achieved by worsening the other value dimensions. Therefore, the Pareto front represents the optimal sub-set of possible service combinations.

First, considering Fig. 2ii, value for the customer is highly correlated with ecological value in the example chosen. This can be explained by the fact that the services for the customer which increase the performance of the equipment inherently also improve the resource efficiency, meaning that more output per resource invested implicitly benefits both the customer's economic goals as well as the ecological goals. Second, as already discussed in Meierhofer et al. (2022), customer value needs to be traded off against provider value, even if the different solutions all reside on the Pareto front (Fig. 2i). Against the background of these two relationships, it becomes evident that also ecological value needs to be traded-off against provider value in the numerical example chosen.

The points A, B, and C labeled in Figs. 2 and 3 all represent Pareto optimal points. For A, the provider creates high value for itself by targeted offerings and retention, but little value for the customer and the ecology by applying no targeted training and just standard service in the expand and stabilize phases. Opposed to this, for B, the provider activates targeted training and the highest service intensity in stabilize (i.e., 5. performance optimization), but no targeted retention, and thus considerably increases V_C and V_{Eco} at the cost of V_p. In contrast to the extreme positions A and B, C represents a balanced Pareto optimal value situation with targeted offering, training and retention activated, and a simple but effective monitoring service in the phase stabilize (intensity 2.). In the example of D, the provider may heuristically assume that providing a good service in phase "Stabilize" and then trying the keep and win back the customers provides optimal value, i.e., neither targeted offering nor training, but remote service during "Stabilize" and targeted retention in "Terminate". However,

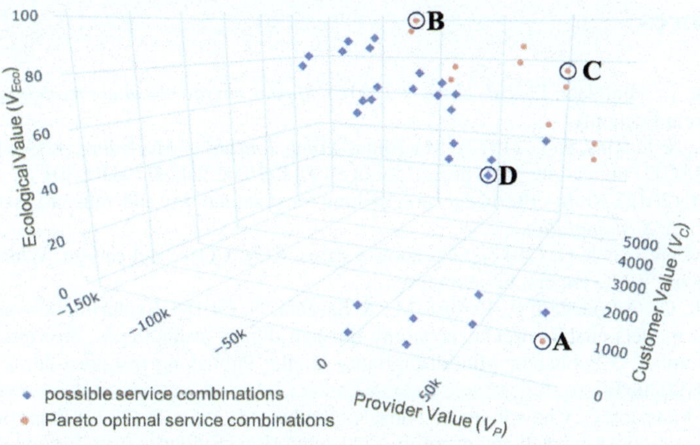

Fig. 3 Value for provider, customer and ecological value with pareto front (red)

Figs. 2 and 3 make evident that point D is below the Pareto front. In particular, the potential for ecological value creation is completely missed out.

5 Conclusions and Recommendation

The paper reveals that the models for economic and ecological value creation can well be combined to an integrated model. The customer lifecycle lends itself to this modelling and enables a differentiated incorporation of the different drivers for benefits and impacts. The combined optimization of both economic and ecological value requires a multi-objective optimization approach since the different value dimensions are not directly comparable.

In its practical implication, the model makes evident that heuristically chosen service constellations often do not provide optimal value creation. Companies need to optimize their services in an informed and systematic way. Additionally, companies become aware that the optimum consists of a set of many possible solutions on the Pareto front, and that there is trade-off between value for service providers, for service customers, and for the ecology.

The concept described in this paper represents a new theoretical approach for an integrated quantitative model of both ecological and economic value in manufacturing service ecosystems. Thus, it extends the existing research literature by a new direction which opens the door for a series of future research questions. Practitioners in industry can apply the model for assessing the design and impact of their services both in economic and ecological dimensions. The resulting set of optimal service constellations enables them to take a conscious strategic decision on their current and future service offering.

References

Achterberg, E., Hinfelaar, J., & Bocken, N. (2016). *Master circular business models with the Value Hill*. Circle Economy.

Bartels, R., & Jenkins, R. L. (1977). Macromarketing. *Journal of Marketing, 41*(4), 17–20. (Publisher: SAGE Publications Inc). https://doi.org/10.1177/002224297704100401

Bertoni, M. (2019). Multi-criteria decision making for sustainability and value assessment in early PSS design. *Sustainability*.

Brander, M., & Davis, G. (2012). Greenhouse gases, CO_2, CO_2e, and carbon: What do all these terms mean. White Papers, Econometrica.

Bressanelli, G., Adrodegari, F., Perona, M., & Saccani, N. (2018). Exploring how usage-focused business models enable circular economy through digital technologies. *Sustainability, 10*(3), 639. (Number: 3 Publisher: Multidisciplinary Digital Publishing Institute). https://doi.org/10.3390/su10030639.

Dey, P. K., Malesios, C., Chowdhury, S., Saha, K., Budhwar, P.,& De, D. (2022). Adoption of circular economy practices in small and medium-sized enterprises: Evidence from Europe. *International Journal of Production Economics, 248*, 108496.

Doni, F., Corvino, A., & Bianchi Martini, S. (2019). Servitization and sustainability actions. Evidence from European manufacturing companies. *Journal of Environmental Management, 234*, 367–378.

Ebeling, J., Friedli, T., Fleisch, E., & Gebauer, H. (2014). Strategies for developing the service business in manufacturing companies. In G. Lay (Ed.), *Servitization in industry* (pp. 229–245). Springer International Publishing. https://doi.org/10.1007/978-3-319-06935-7_14.

Ecoinvent. (n.d.). SimaPro Ecoinvent 3.6 Database (Tech. Rep.). Retrieved 2022–09-09, from https://ecoinvent.org/the-ecoinvent-database/

Jasiulewicz Kaczmarek, M., & Gola, A. (2019). Maintenance 4.0 technologies for sustainable manufacturing—an overview. *IFACPapersOnLine, 52*(10), 91–96.

Kowalkowski, C., & Ulaga, W. (2017). *Service strategy in action: A practical guide for growing your B2B service and solution business*. Service Strategy Press.

Kühl, C., Tjahjono, B., Bourlakis, M., & Aktas, E. (2018). Implementation of Circular Economy principles in PSS operations. *Procedia CIRP, 73*, 124–129.

Meierhofer, J., Benedech, R., & Heitz, C. (2022). On the value of data: Multi-objective maximization of value creation in data-driven industrial services. In *2022 9th swiss conference on data science (SDS)*. IEEE Xplore.

Meierhofer, J., Heitz, C., & Hannich, F. (2021). Optimizing service value creation with smart, connected products. In *Proceedings of the 2021 naples forum on service* (p. 13). Capri.

Rapaccini, M., & Adrodegari, F. (2022). Conceptualizing customer value in data-driven services and smart PSS. *Computers in Industry, 137*, 103607.

Schüritz, R., Farrell, K., Wixom, B. W., & Satzger, G. (2019). Value co-creation in data-driven services: Towards a deeper understanding of the joint sphere. *ICIS 2019 Proceedings*.

Tecchio, P., McAlister, C., Mathieux, F., & Ardente, F. (2017). In search of standards to support circularity in product policies: A systematic approach. *Journal of Cleaner Production, 168*, 1533–1546.

Tukker, A. (2015). Product services for a resource-efficient and circular economy—a review. *Journal of Cleaner Production, 97*, 76–91.

Vendrell-Herrero, F., Vaillant, Y., Bustinza, O. F., & Lafuente, E. (2021). Product lifespan: the missing link in servitization. *Production Planning and Control*, 1–17. (Publisher: Taylor & Francis eprint). https://doi.org/10.1080/09537287.2020.1867773

Zheng, T., Ardolino, M., Bacchetti, A., & Perona, M. (2020). The applications of Industry 4.0 technologies in manufacturing context: a systematic literature review. *International Journal of Production Research*, 1–33. (Publisher: Taylor & Francis eprint). https://doi.org/10.1080/00207543.2020.1824085

A Sustainable Human-Smart Service Product Interaction: Tackling User Attrition and eWaste for the Triple Bottom Line

Cecilia Lee, Atul Gupta, and Utpal Mangla

Abstract This paper takes a design research approach to explore the challenge of user attrition in the consumer wearables sector and its impact on eWaste which is a growing concern for many countries. Previous research demonstrated the lack of value-in-use realised by users as the key driver of user attrition. The abandoned smart service products then become the source of mounting eWaste which creates environmental pollution and puts human health at risk. This study hypothesises that a sustainable interaction between user and consumer wearables will reduce the user attrition rate, which will, as a result, decrease the amount of eWaste. This hypothesis is examined through three case studies that showcase the past, the present, and the future of the consumer wearables sector. The study contributions are as follows. First, it takes a zoom-in and a zoom-out approach to understand how the challenge of user attrition at the micro-level leads to the challenge of environmental sustainability at the macro-level. Secondly, this study introduces how the design research community uses case studies as a method in design research to the non-design research community. Lastly, three propositions introduced in this research could inform the foundation of a sustainable human-smart service product interaction framework.

Keywords Smart service product · User engagement · User attrition · Value co-creation · eWaste

C. Lee (✉)
Royal College of Art, London, UK
e-mail: cecilia.lee@network.rca.ac.uk

A. Gupta
Sun Life, Toronto, Canada
e-mail: aguptab@gmail.com; atul.gupta@sunlife.com

U. Mangla
IBM, Toronto, Canada
e-mail: utpal.mangla@ca.ibm.com

J. Meierhofer et al. (eds.), *Smart Services Summit*, Progress in IS,
https://doi.org/10.1007/978-3-031-36698-7_4

1 Introduction

In Industry 4.0 era, smart services have become ubiquitous. Whether it be fitness wearables or digital voice assistants, our lives in a smart service economy so-called Industry 4.0 are bombarded by the influx of smart service products enabled by emerging technologies, such as artificial intelligence (AI) and the Internet of Things (IoT), to name a few. Consumer IoT products, especially fitness trackers and smartwatches have become mainstream nowadays. Although their user adoption has grown consistently over the past years, the lack of user engagement that leads to attrition has been an ongoing challenge for many consumer wearables makers (Ashfaq et al., 2020).

User attrition does not happen overnight. The lack of user engagement ultimately leads to user attrition. Previous studies that examined the growing user abandonment of consumer wearables found that the lack of user value is the key driver of the abandonment (Ashfaq et al., 2020; Clawson et al., 2015; Lazar et al., 2015). These studies focused on a micro-level problem, the lack of user value, which results in a lack of user engagement that leads to attrition. However, the lack of user value that leads to user attrition can potentially cause a much bigger problem that affects not only end users, but also the environment, society, and businesses at large. The recent blog post published by Microsoft calls for the public attention towards carbon emissions required for smart service product (SSP) production. For example, the author of the blog post states that the average smartphone needs approximately 55 kg of CO_2 for manufacturing, and smartphone manufacturing accounts for up to 95% of the annual carbon footprint of the phone (Manne, 2020). This is an alarming rate, especially when the climate change crisis has become prominent in many parts of the world these days. Similarly, according to the Global eWaste Monitor (2020), eWaste is expected to reach approximately 74.8 Mt by 2030.

This article aims to explore how SSP makers can better enable the value co-creation process between user and SSP in order to enhance user engagement, retain users, and fulfil the triple bottom line—the environmental, social, and business sustainability. We hypothesise that a seamless value co-creation between user and SSP will lead to a virtuous cycle where user's value-in-use is realised, which also contributes to the sustainability of the environment, society, and businesses. We take a design research approach and adopt a case study as a method to analyse the challenge of user attrition in consumer wearables sector through the lens of past and present. We then explore how emerging opportunities enabled by the advancement of technologies and better insights on user needs would likely evolve the challenge of user engagement and user attrition in the future. We have specifically chosen the case studies of the consumer wearables sector, as fitness trackers and smartwatches have reached the mainstream and have been experiencing user attrition consistently (Windasari et al., 2021).

2 Human-Smart Service Product Interaction in a Smart Service Economy

The concept of an SSP is derived from the notion of service-dominant logic (SDL) thinking, which views the fundamental basis of economic and social exchanges as a service. SDL was first introduced in 2004 by Vargo and Lusch who brought a paradigmatic shift in thinking to a society predominated by goods-dominant logic (GDL) in which customer (or user) value is seen as embedded in physical product itself. SDL thus offers an alternative thinking towards the concept of value. In an SDL world, value is not realised until it is actively co-created by all the participating actors, including both service providers and beneficiaries, and value outcome is uniquely and phenomenologically determined by the beneficiary (Vargo & Lush, 2004, 2008).

Many consumer wearables such as fitness trackers and smartwatches have experienced a consistent user adoption since 2014, but their bulky design, a lack of battery power, and many frictions in the user journey, such as multiple steps to set the device up for a daily use, had put users off. In the year 2021, many aspects of SSPs that were causing the users to abandon SSPs over the past few years seem to have resolved.

However, the challenge of user attrition remains a concern (Windasari et al., 2021). Also, another growing concern is eWaste mounting from fitness trackers. According to the sales data from the electronics retailer in the UK, Currys PC, the sales of fitness trackers have grown 45% year on year, but only 13% of those who have purchased them have chosen to recycle them, while 11% of them simply put them into the bin (Parsons, 2021). Whether fitness trackers are disposed because users want a new model or no longer need it, it is becoming a growing concern, as fitness trackers and smartwatches are not easy to be dismantled and recycled, according to the American Chemical Society (2021). The UK government recently published the report on the Environmental Audit Committee to ban smart service providers, such as Amazon and Apple, from intentionally shortening the life cycle of SSPs although this may help reduce eWaste from users replacing their old gadgets with new model frequently, it will not be of a great help if users dispose fitness trackers because simply they no longer need it.

The value co-creation process between user and consumer wearables requires a careful reassessment; therefore, users can realise their value-in-use and are encouraged to engage with their fitness trackers and smartwatches in a long run. Figure 1 illustrates a virtuous cycle of a smart service economy where SSP makers play a pivotal role.

Figure 1 shows how a sustainable interaction between a user and an SSP can result in less eWaste, as users will engage with an SSP in a long run and therefore SSP makers will likely focus on a new service development rather than launching brand new devices, which is a significant source of environmental pollution and human health risks. New service offerings will enable smart service product makers to better facilitate the value co-creation process between user and SSP that may lead to the fulfilment of the environment, social, and business sustainability. In doing so,

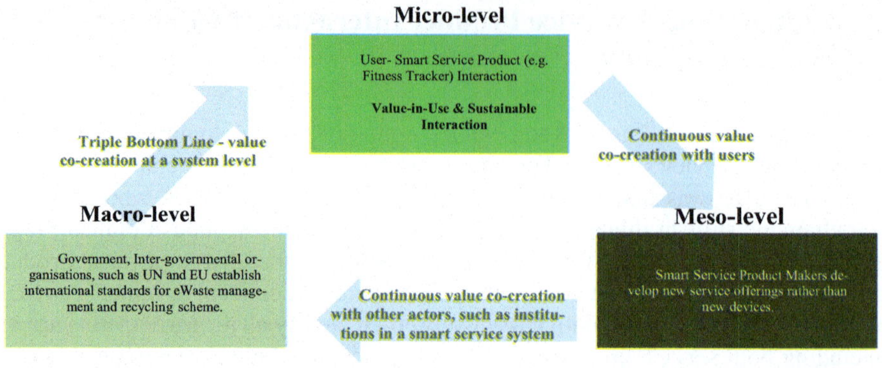

Fig. 1 A circular economy for smart service system

a smart service system generates value for the mutual benefits of actors in the system and further strengthens the sustainability of the system.

3 Research Question

Although the extant literature have explored the challenge of engaging and retaining the users of consumer wearables in a long run, these studies have not yet examined how smart service providers, such as consumer wearables makers, could better enable the value co-creation process between users and consumer wearables and ultimately reduce the user attrition rate and eWaste. In response to this gap in knowledge observed in the existing literature, we inquire into the research question as follows:

How could we better enable the value co-creation process between user and consumer wearables to enhance user engagement, retain users in a long run, and reduce eWaste that contributes to the environmental, social, and business sustainability? We take a design research approach and adopt a case study as a method to explore this research question.

4 Methodology

4.1 Case Study as a Method in Design Research

Design research does not share an epistemological and ontological view of scientific research originating from natural science; as a result, design research had often been criticised for the lack of rigour, as its approach does not conform to the rules and procedures defined by natural science. Similarly, a case study which has often been

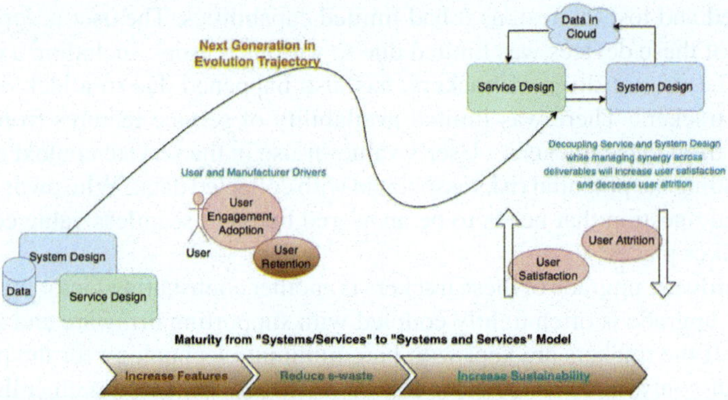

Fig. 2 The potential evolution trajectory of consumer wearables

used as a research method in social science has also been the target of criticism, as a case study lacks external validity in positivists' view. Teegavarapu and Summers (2008) claim that design research and case study share many similarities, especially in the aspect that historically they have not had a systematic process in place to conduct research. They also argue that although design researchers have not explicitly stated in their work, many design research has adopted a case study as a method in their study, which is supported by increased publications that used a case study as a method in the *Journal of Engineering Design.*

With some guidance from existing literature in social science and design research, this paper uses a case study as a method to examine the research question posed in a previous section. We chose three case studies, following the potential evolution trajectory of the consumer wearables sector in Fig. 2 to closely inquire into the challenge of user engagement and attrition observed in the past and present world and explore how emerging opportunities enabled by the advancement of technologies and a better understanding of user needs and desires would likely evolve the challenge of user engagement and user attrition in the future. More specifically, the first case looks at fitness trackers that have reached the market maturity but are still facing an ongoing challenge of user engagement and attrition. The second case study explores the current state of consumer wearables that lacks a wide adoption and engagement. The last case study discusses the service enablement for healthcare providers, which is still at the development stage within research labs of higher education institutions and software companies.

4.2 *Case Study #1: Fitness and Vitals Trackers*

The first generation of fitness and vital trackers were adopted by consumers by virtue of fashion, and soon after users' interest in them started shrinking. The users became disengaged and lost interest, as it had limited capabilities. The user acceptance and adoption of these devices was limited due to various reasons, including a premature launch. The user attrition of trackers, has also happened due to a lack of services for these trackers. There was limited availability of service features from the data collected from these trackers. User's value-in-use in the service context starts with understanding the potential risks associated with collected data. 'Who owns the data?' is a crucial question that needs to be answered before a seamless value co-creation process takes place.

The hardware upgrade of these trackers is another contributing factor for user attrition. This upgrade is often tightly coupled with supporting firmware and embedded software. If the trackers are kept with bare minimum technology for the purpose of data capture only, the service layer can be constantly enhanced with collected data stored and analysed in cloud and edge layers. This can eventually build a more fine-grain service layer which can inform more sophisticated service solution. In doing so, it removes the need for constantly building solution within the tracker hardware.

4.3 Case Study #2: Wearables

The next generation of wearables such as stretchable sensors, synthetic skin, (Sheedy, 2021) biodegradable sensors and wearables (Ghosh et al., 2021), can truly revolutionise the growing e-waste challenges. As mentioned, these wearables are next generation and are still on the path towards a wide market adoption. The potential key barrier to user adoption and engagement would be users' fear of the use of wearables, since users need to stick them on their skin, or have it in a close contact with their body. Since, this next generation of wearables are still in labs, there is an opportunity to consider decoupling these devices for the sole purpose of 'data-capture'. The data capture capability from the wearables, human body sensors, and vicinity sensors should be the only focus of these hardware technologies. The service layer, edge computing, and perimeter IoT's should be responsible for compute, analysis and running AI/ML algorithms. The AI guidance from the service layer should be brought back to the users via traditional mechanisms, such as phones, tablets and other household technology enabled gadgets.

A new design approach that integrates service design could help enhance user adoption and user engagement in this category, which, as a result, could reduce the user attrition rate and the amount of e-waste from consumer wearables. In doing so, it will move us towards more focused, decoupled, and reusable hardware.

4.4 *Case Study #3: Service Enablement for Healthcare Providers*

The above two examples are either in users' hands or soon to be in users' hands, but this example of 'service enablement' for healthcare providers has yet to become a reality. The key concept of this example is decoupling between system design and service design. Decoupling starts with use of federated data from many different sources, leaving the system design maturity to itself. The hardware or system design maturity is often combined with service design maturity, but decoupling them is needed for maturing service enablement, because many technology vendors attribute their success to the commercial success of their hardware offerings. This widely held belief amongst vendors often leads to thc lack of motivation for building collaboration to develop a product service ecosystem.

One of the sustainability issues caused by user attrition of wearables is due to a lack of new approaches in service enablement. The smart services and user interaction are needed at two perimeters—one is where 'data origination' from the sensors/wearables takes place and the other is when 'feedback and guidance' to the users are provided. The service enablement for healthcare professionals and providers should be trusted by end-users (patients). Maturity of services with time will make it sharable and sustainable. The sophisticated AI solutions can fulfill the healthcare capabilities with hardware and services decoupled, reduced interdependencies and developing synergised ecosystem.

5 Result and Discussion

We analysed three case studies of the consumer wearables sector, following the potential evolution trajectory of this sector in Fig. 2 to inquire into the challenge of user engagement, user attrition, and eWaste. The analysis outputs can be found in Table 1.

Our analysis in Table 1 helped us to develop the three propositions that could become the foundation of a sustainable human-smart service product interaction framework in Table 2.

6 Study Contributions

The contributions of this study are three-fold. First, existing literature in consumer wearables that examined the user engagement and attrition have not yet examined its spill-over effect on mounting eWaste, which is a growing concern for many countries. Our study explored the micro-level problem—the lack of user engagement and attrition—and how the improvement of these problems can lead to the reduction of

Table 1 Common challenges and opportunities for user engagement, attrition, and eWaste

	Past	Present	Future
	Fitness and vitals tracker	Wearables	Service enablement for healthcare providers
User engagement	These trackers were first generation and limited maturity with data-capture	Generic adoption can be used across platform due to inter-connectivity and users can be engaged in non-intrusive way	Use existing devices, trackers in the ecosystem, user engagement pre-exists
User attrition	High attrition observed as too many versions came too often	With emphasis on software layer and inter-connectivity, attrition rate is expected to be low	Not applicable or very minimal
Data usability/re-usability	Almost no data re-usability, and data usability was proprietary	Generic re-usable data layer can evolve of it	Service and analytics layer matures to attain new levels of excellence

Table 2 The propositions for a sustainable human-smart service product interaction

Proposition 1	SSP makers must consider keeping bare minimum technology to maximise service layer that can co-create value with end users
Proposition 2	SSP makers should be able to better facilitate the service enablement for value co-creation with end users
Proposition 3	SSP makers need to better leverage service design and system design approaches to enable value co-creation, not only at the micro-level of the system, but also at the connected ecosystem level

eWaste at the macro-level of the smart service system. Secondly, we took a design research approach and adopted a case study as a method. Our study introduces how a case study can be leveraged in design research to the non-design research community. Lastly, our analysis put forward three propositions that could inform the foundation for a sustainable human-smart service product interaction framework.

References

American Chemical Society. (2021, August 4). *Dissolvable smartwatch makes for easier electronics recycling*. https://www.sciencedaily.com/releases/2021/08/210804123515.htm

Ashfaq, M., Yun, J., Yu, S., & Loureiro, S. M. C. (2020). I, Chatbot: Modelling the determinants of users' satisfaction and continuance intention of AI-powered service agents. *Telematics and Informatics, 54*(2020), 101–473. https://doi.org/10.1016/j.tele.2020.101473

Clawson, J., Pater, J. A., Miller, A. D., Mynatt, E. D., & Mamykina, L. (2015, September 7–11). *No longer wearing: Investigating the abandonment of personal health-tracking technologies on craigslist* [Conference session]. UBICOMP 2015, Osaka, Japan. https://doi.org/10.1145/2750858.2807554

Ghosh, S. J., Park, J., Na, S., Kim, M. P., & Ko, H. (2021). A fully biodegradable ferroelectric skin sensor from edible porcine skin gelatine. *Advanced Science, 2021*(8), 1–11. https://doi.org/10.1002/advs.202005010

Global E-Waste Monitor. (2020). *Global E-Waste Monitor 2020*. https://www.itu.int/en/ITU-D/Environment/Pages/Spotlight/Global-Ewaste-Monitor-2020.aspx

Lazar, A., Koehler, C., Tanenbaum, J., & Nguyen, D. H. (2015, September 7–11). *Why we use and abandon smart devices* [Conference session]. UBICOMP 2015, Osaka, Japan. https://doi.org/10.1145/2750858.2804288

Manne, S. (2020). *Examining the carbon footprint of devices*. Microsoft Sustainable Software. https://devblogs.microsoft.com/sustainable-software/examining-the-carbon-footprint-of-devices/

Parsons, J. (2021, January 1). *Home working sparks tech boom and e-waste concerns*. https://metro.co.uk/2021/01/01/home-working-sparks-tech-boom-and-e-waste-concerns-13833275/

Sheedy, C. (2021, July 13). *This electronics engineer is giving sensation to synthetic skin*. https://createdigital.org.au/electronics-engineer-giving-sensation-to-synthetic-skin/

Teegavarapu, S., & Summers, J. D. (2008, August 3–6). *Case study method for design research: A justification* [Conference Paper]. ASME2008 International Design Engineering Technical Conference and Computers and Information in Engineering, Brooklyn, New York, USA. https://doi.org/10.1115/DETC2008-49980

Vargo, S. L., & Lush, R. F. (2004). Evolving to a new dominant logic for marketing. *Journal of Marketing, 68*, 1–17.

Vargo, S. L., & Lusch, R. F. (2008). Service-dominant logic: Continuing the evolution. *Journal of Academy of Marketing Science, 36*, 1–10.

Windasari, N. A., Lin, F., & Kato-Lin, Y. (2021). Continued use of wearable fitness technology: A value co-creation perspective. *International Journal of Information Management, 57*(2021), 1–15. https://doi.org/10.1016/j.ijinfomgt.2020.102292

Conceptual Ecosystems to Support the Development of Sustainable Business Models for a Capital Equipment Manufacturer

Daniel Wörner, Manuel Ritter, Shaun West, and Thomas Friedli

Abstract Improving sustainability concerns many stakeholders: customers, shareholders, employees, and regulators. The manufacturing industry is a significant factor as production continually depletes finite resources. Without a shift to more sustainable business models (i.e. those based on the triple bottom line), there is a risk of continued depletion of resources. A common finding from prior work is the need to act within an ecosystem approach to support the development of a circular economy that considers the multiple interactions and dynamic recombination of resources that lead to value co-creation. This paper investigates the interplay of ecosystem actors and builds conceptual habitats to improve sustainability. It considers the strategic meso and macro levels of the ecosystem to provide detailed insights into the behaviors and motivations of different actors. Critical bilateral and multilateral interdependences among the ecosystem actors are highlighted. The presence of digitalization provides the necessary institutional arrangements for interactions between diverse actors. This builds upon the work of West et al. (Practices and tools for servitization. Springer, pp. 363–385, 2018), by creating multiple conceptual future state ecosystems, identifying gaps within the ecosystems, and providing a description of the actors. The study suggests that an extended model is considered for developing and operationalizing sustainability-based business models for capital equipment. The limitations of this study reveal the requirement to validate and cross-check the suggestions in a real case example with a broader scope for comprehensive qualitative and quantitative data collection.

D. Wörner (✉) · M. Ritter · T. Friedli
University of St. Gallen, 9000 St. Gallen, Switzerland
e-mail: daniel.woerner@unisg.ch

M. Ritter
e-mail: manuel.ritter@unisg.ch

T. Friedli
e-mail: thomas.friedli@unisg.ch

S. West
Lucerne University of Applied Sciences and Arts, 6048 Horw, Switzerland
e-mail: shaun.west@hslu.ch

J. Meierhofer et al. (eds.), *Smart Services Summit*, Progress in IS,
https://doi.org/10.1007/978-3-031-36698-7_5

1 Introduction

Actors' interactions build an ecosystem based on interdependencies (Jacobides et al., 2018). Ecosystem actor characteristics and business models define environments influential to the viability of capital equipment manufacturers. Deriving from the triple bottom line as a concept to frame sustainability and business conduct, and leaning on a circular economy perspective, equipment manufacturers' current business models lack inclusion and consideration of their ecosystem influences (Hauschild et al., 2017). Social, economic, and environmental aspects of the triple bottom line are insufficiently covered by ill-equipped business models. In the transition and development of a sustainable business model (SBM) for equipment manufacturers, understanding current stakeholder interdependencies exhibit challenges to overcome and opportunities to realize (Stubbs & Cocklin, 2008). Bringing SBMs into operation demands an innovation-driven shift in value creation, delivery, and capture (Bocken et al., 2014). Interlinked activities of equipment manufacturers and ecosystem actors along the equipment life cycle build the source for value generation. While some strategies such as repair are standardized for many equipment manufacturers, other strategies, such as recycling equipment, experience hesitations or restrictions (de Pádua Pieroni et al., 2018). Insufficient knowledge on the realization of sustainability-oriented and environmentally beneficial strategies reveals a gap between conceptualization and operationalization.

To clarify the misalignment of ecosystem actors' behaviors and their influences on an equipment manufacturer's SBM development, the study describes observed characteristics and synthesizes them towards role identifications. In doing so, the paper answers the question: "How are ecosystem actors interrelated with the equipment manufacturer's activities along the equipment life cycle and what conditions lead towards a sustainable business model?"

2 Concept and Theory

The triple bottom line is promoting the consideration of the social and environmental layer, additionally to the firm-centric economic perspective, as emphasized by the Brundtland Report (1987). Advancing environmental progress and reaching sustainability along these dimensions predominantly works with business agendas that are constituted by choice (Elkington, 2004). The financial markets expect companies to act towards profitability, neglecting the other two layers. Hauschild et al. (2017) suggest a novel perspective on the triple bottom line by arranging the layers dyadically dependent, with manufacturing industry at the center. Equipment manufacturers typically deplete natural resources along the supply chain, which negatively impacts the environment, i.e., "earth's life support system" (Rockström, 2015). Actors outside a company, such as society or regulation, incentivize changes of behavior, towards the inclusion of the social and environmental layers, by requiring the firm to reach certain goals. Realizing these opportunities justifies the operating viability of an equipment manufacturer.

Traditional business models are based on cradle-to-grave thinking and linear production patterns (Lieder & Rashid, 2016). A common approach to operating more environmentally sustainably i.e., moving from cradle-to-grave to cradle-to-cradle and being aware of the circular economy, is "circular strategies" (MacArthur, 2013). Bocken et al. (2016) emphasize that these strategies aim to narrow (optimize natural resources, e.g., redesign to use less material), slow (extend the product's life, e.g., repair), or close (use the product after its use phase, e.g., recycling) sections of the equipment life cycle. Equipment manufacturers operate some measures, while others build upon collaborative value creation, reflecting multi-stakeholder management and business ecosystems, as showcased by Stubbs and Cocklin (2008). Aiming towards sustainability, an equipment manufacturer manages the value co-creating stakeholders along the life cycle and sustainable strategies, matching its business model (Romero & Molina, 2011).

Following the definition of an ecosystem by Jacobides et al. (2018, p. 2264): "*An ecosystem is a set of actors with varying degrees of multilateral, nongeneric complementarities that are not fully hierarchically controlled*", actor interrelations build an ecosystem. Equipment manufacturers must be aware of their ecosystem and its actors, to foster mutual value creation (Lozano, 2018). An understanding of value creation and exchange of perceived services from manufactured goods yields the service-dominant logic coined by Vargo and Lusch (2004). Actors co-evolve their capabilities and functions based on their position in the ecosystem (Moore, 2006). Due to these continuous exchanges, the nature of ecosystems is dynamic, yielding scrutiny of an actor's existing business structures and sustainability approaches (Liu & Stephens, 2019).

The values exchanged in the ecosystems are created from company-specific business models that generate added value. Business models are important to guide a company to conduct its business to competitive advantage (Bocken et al., 2014).

A generic business model concept consists of: (1) value proposition, (2) supply chain, and (3) customer interface (Boons & Lüdeke-Freund, 2013). While the supply chain and customer interface aspects are captured in the perception of ecosystems, the value proposition is inherent to a company's offerings. Reflecting on the importance of sustainability (Brundtland, 1987), Geissdoerfer et al. (2018) exhibit value proposition, value creation and delivery, and value capture for the economic, environmental, and social sustainability dimensions toward sustainable and eventually circular business models. A definition for SBMs is provided by Lüdeke-Freund (2010, p. 23): "*A business model that creates competitive advantage through superior customer value and contributes to a sustainable development of the company and society …*".

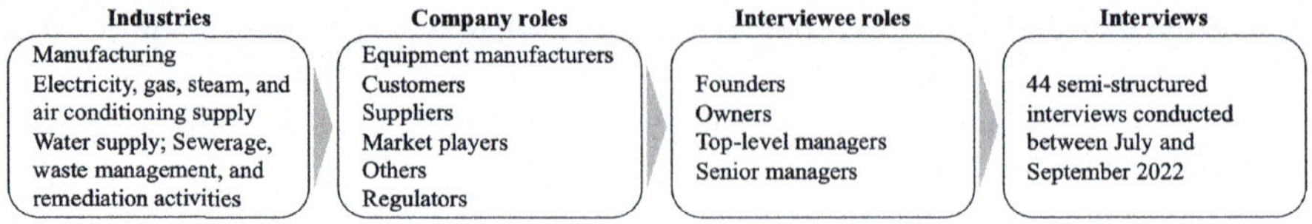

Fig. 1 Selection of interviewees along a set of industries with diverse contextual backgrounds

3 Methodology

The study investigates in an explorative manner and applies a qualitative research design (Yin, 1994). An inductive research approach leans on the suggestions by Gioia et al. (2012), and aims to develop a theory from the data (Strauss & Corbin, 1994).

The qualitative data was aggregated from cross-industrial semi-structured interviews examining status quo business models and ecosystem interdependencies. Selected industries are classified according to the Nomenclature of Economic Activities (European Commission, 2022). Company roles were carefully balanced to get a diverse set of opinions. Emphasis was put on anonymity to ensure open discussions, potentially revealing deeper insights. The qualitative data collection is visualized in Fig. 1.

First, the data was analyzed for statements implying interdependent relationships with stakeholders along the equipment life cycle (1st order codes). Second, these statements were mapped to inclusive information properties across the ecosystems (2nd order themes). These preliminary findings were clustered under umbrella terms valid across the defined ecosystems (aggregate dimensions). The data structure can be shared upon request. Finally, the interrelationships of the ecosystem actors were examined.

4 Findings

The analysis assessed the qualitative data on present challenges across ecosystems. The identification of challenges sheds light on the motivations, incentivization and behaviors of the ecosystem actors. Classifications and clusters of the qualitative data eventually yielded five common challenges across industries.

The consolidated findings offer comparative parameters to examine interrelationships among ecosystem actors from the equipment manufacturer's vantage point. Based on these challenges, the examination identifies the relationships among actors. Subsequently, Table 1 highlights bi- and multi-lateral interdependencies with respect to "Culture", "Operation", "Information", "Market", and "Legislation". The categorization of the inductively assessed challenges has overlapping traits with the sectors of the external environment exhibited by Stead and Stead (2014), taking a holistic vantage point.

Table 1 Interdependencies among ecosystem actors according to common challenges

Challenges in conceptual habitats		Equipment manufacturer				
		Customer	Supplier	Market player	Third-party	Regulator
Culture challenges	Financially-driven management incentivization	● —	— ●		●	
	Misalignment of cultural mindsets	●				
	Mismatch of strategic roadmaps to sustainability	● —	— ● —	— ● —	— ●	
Operation challenges	Limitations due to product properties	● —	— ●		● —	— ●
	Sale-orientation of traditional business models	● —	— ● —	— ● —	— ●	
	Immaturity of technological implementation	● —	— ● —	— ● —	— ●	
Information challenges	Restricted data availability	●				●
	Lack of information availability	● —	— ●			
Market challenges	Dominance of cost-efficiency	● —	— ●		●	
	Lack of life cycle thinking	●		●		●
	Scarcity of human resources	●			●	
Legislation challenges	Hindering regulatory framework	●		● —	— ●	●
	Lack of transparency along the supply chain	● —	— ● —	— ● —	— ● —	— ●
Shared value attributes with the equipment manufacturer		Value creation Value capture	Value delivery	Value creation Value delivery	Value creation Value delivery	(Value creation) (Value delivery)

In the table below, Customers receive the equipment manufacturer's value (co-) creation and make sense of the value captured in the business ecosystem. They are represented in all the challenges uncovered, which can turn into opportunities leading to more potential value creation. Market challenges are more prevalent for customers than for the third-party actors. Suppliers deliver virgin materials and semi-finished products to equipment manufacturers. Each actor in the supply chain contributes to value delivery and influences the availability of environmentally sound resources. Most identified hurdles are Operation challenges, such as limited availability of specific resources, and Culture challenges occur when aligning sustainability agendas along the supply chain.

'Market players' are firms active in similar markets and can be competitors. For the equipment manufacturer, they share the challenges of translating sustainability into business conduct, which cluster in Operation and Culture challenges. These efforts create more environmentally friendly operations or apply life cycle thinking. Market players and the equipment manufacturer can share similar visions and goals, making them prone to competition but enhancing opportunities to co-create and deliver value.

The 'Third-party' category includes third-party services and providers in the equipment manufacturer's post-manufacturing stages of the supply chain, e.g., spare parts dealers, workshops, distributors, etc. Due to shared business model characteristics, interdependencies show in Operation challenges as well as Market challenges. Third-parties can deliver value as key partners or create value by sustaining distribution channels.

Regulators build the legislative framework that guides a company's behavior to have and sustain legitimacy in the market. They set holistic goals for sustainability, representing boundary conditions of the people and the planet. As Legislation challenges, these targets can become a hurdle for equipment manufacturers, e.g., in inherently energy-intense sectors such as power. Regulators also pose conditions in terms of data transparency and availability, in Information challenges. Their values indirectly influence the equipment manufacturer company's key activities and value propositions.

5 Discussion

The findings exhibited challenges and interdependences among ecosystem actors toward sustainable business models. Traditional business models, which tend towards linear and depleting production patterns, reach a cap for generating additional value and often focus one-dimensionally on the relationship with their stakeholders. A firm-centric orientation can result in unsustainable business conduct. In the light of social, economic, and environmental sustainability, value is (co-)created in ecosystems. This requires equipment manufacturers to adopt ecosystem thinking, and consider the influences ecosystem actors have. Such influences manifest as challenges or as benefits and can change according to the dynamics of actors within ecosystems. Seeking actions that align with the business agendas of the equipment manufacturer and at least another actor in the system can lead to value co-creation opportunities. Kohtamäki and Rajala (2016) describe the state of value co-creation and coproduction of value propositions as "... *vantage point from which to comprehend the inter-organizational, dynamic, and systems-oriented view of value creation.*" Instrumentalizing this perspective of value creation includes a set of actors and requires an understanding of their interdependencies to evaluate the validity and feasibility of co-creation. Equipment manufacturers should expose themselves to co-creation opportunities as a pathway to sustainable market competitiveness. In addition, the transition from traditional (linear) business models to sustainable business models includes an extended scope of ecosystem actors as the consideration of novel practices (e.g., recycling activities) demands the inclusion of new ecosystem actors (e.g., partnerships with scrap dealers). These loops of dynamic recombination evolve alongside challenges and emerging opportunities.

Major opportunities can be deduced from manufacturers' relationships with customers and vice versa. As mentioned earlier, customers are central to creating and capturing value. This observation is in line with Lusch and Vargo (2006, p. 284): "*The customer [beneficiary] is always a co-creator of value*". Consequently, products and services cannot be developed and sold without the involvement of the customer and value creation stays uncaptured. Initiating campaigns and investing in efforts to shift and align the cultural mindset can support the equipment manufacturer to offer economically viable and environmentally sustainable services. Also, creating data-sharing interfaces and communication and information streamlining technologies that respect the customer's privacy can improve the equipment manufacturer's value proposition. It can also improve products and services, which eventually benefits the customer base.

Considerable potential lies in the sourcing and mining of sustainable materials such as bio-based materials or non-hazardous substitutes. The equipment manufacturer's willingness to pay a premium should meet the supplier's requirements and pursue the sustainability agenda. A driver to converting value delivery to value capture is transparency along the supply chain, which the supplier and the equipment manufacturer may approach together via technological interfaces or disclosed agreements.

With Kohtamäki and Rajala's (2016) description of value co-creation, market players are vital to agree on collaboration beneficial to both or more parties. Aligning with the cultural mindset and the sustainability agenda can support them. Depending on the size of the companies, consortia or associations may be a pathway to discuss the regulatory framework with institutions and initiate change on a legislative level.

Third-party product and service providers should be seen by the equipment manufacturer as integrative ecosystem actors. For example, a specialized overhaul workshop maintains a set of highly skilled human resources, whereas the equipment manufacturer falls short due to a different strategic focus. Operationalizing certain sustainable practices, e.g., remanufacturing, could vastly profit from practice-oriented cooperation.

Regulators' sustainability goals can define an equipment manufacturer's ability to operate. These requirements and guidance are goals for the whole market. Against that background, an idealized capital equipment manufacturer must adapt to its ecosystem setting. Core competencies should align with cultural mindsets fostering sustainability, products offering life-extending approaches, processes transparent along the holistic supply chain, life cycle thinking through the stakeholder network, and adherence with the regulatory framework. Assessed challenges should be viewed as opportunities to enhance current value creation and uncover additional potential. Operationalizing these opportunities will enable an equipment manufacturer to gain a competitive advantage and drive added value along the triple bottom line.

Despite the presence of a company's strategic sustainability trajectory and intention, common challenges can impede its realization and major transformative business model evolution. Understanding and assessing specific hurdles in the stakeholder network can improve interdependencies and mutual value creation by transparently organizing and managing a common issue. Optimally, the actors agree on competence distribution, so the most suitable and skilled actor leads operations to counter the common challenge.

These common challenges cover a broad spectrum relevant to equipment manufacturers yet are not complete. Future studies may conduct a firm-specific case study to identify challenges that currently fall out of scope. The findings would vary according to the industry investigated. Thus, the study would be true for a specific industry. Another expansion for further studies would be a quantitative assessment of the challenges, where feasible. The lack of weighting in the proposed challenges is a limitation that could be improved by applying variable factors based on quantitative data.

6 Conclusion

Capital equipment manufacturers are exposed to a set of challenges throughout the ecosystem. These hurdles require companies to find novel ways to create, deliver, and capture value. Transitioning from traditional business models to sustainable business requires an ecosystem approach rather than firm-level improvements. The study identified five common challenges to understanding the status quo of ecosystems: challenges in culture, operation, information, market, and legislation. Confronting these with relevant ecosystem actors yielded a set of interdependencies that demand consideration for value-adding processes. With the examination of a considerable set of interviews with actors in the industrial sphere, this study contributes to research and practice by revealing unnoticed shared challenges occurring in multiple industries and clarifying ecosystem interdependencies influencing the value generation potential of current and future business models. The study has limitations in the selection of investigated ecosystem actors that should be extended in future studies. Further, evaluating the effect on value generation by overcoming the challenges would yield potential incentivization to capital equipment manufacturers, which could be researched in a case study analysis.

References

Bocken, N. M., De Pauw, I., Bakker, C., & Van Der Grinten, B. (2016). Product design and business model strategies for a circular economy. *Journal of Industrial and Production Engineering, 33*(5), 308–320.

Bocken, N. M., Short, S. W., Rana, P., & Evans, S. (2014). A literature and practice review to develop sustainable business model archetypes. *Journal of Cleaner Production, 65*, 42–56.

Boons, F., & Lüdeke-Freund, F. (2013). Business models for sustainable innovation: State-of-the-art and steps towards a research agenda. *Journal of Cleaner Production, 45*, 9–19.

Brundtland, G. H. (1987). Our common future—Call for action. *Environmental Conservation, 14*(4), 291–294.

de Pádua Pieroni, M., Blomsma, F., McAloone, T. C., & Pigosso, D. C. (2018). Enabling circular strategies with different types of product/service-systems. *Procedia CIRP, 73*, 179–184.

Elkington, J. (2004). Enter the triple bottom line. In *The triple bottom line: Does it all add up?* (1st ed., pp. 1–16). Routledge.

European Commission. (2022). *Complete list of all NACE Code.* NACE Codes. https://nacev2.com/en

Geissdoerfer, M., Vladimirova, D., & Evans, S. (2018). Sustainable business model innovation: A review. *Journal of Cleaner Production, 198*, 401–416.

Gioia, D. A., Corley, K. G., & Hamilton, A. L. (2012). Seeking qualitative rigor in inductive research: Notes on the Gioia methodology. *Organizational Research Methods, 16*(1), 15–31.

Hauschild, M. Z., Herrmann, C., & Kara, S. (2017). An integrated framework for life cycle engineering. *Procedia CIRP, 61*, 2–9.

Jacobides, M. G., Cennamo, C., & Gawer, A. (2018). Towards a theory of ecosystems. *Strategic Management Journal, 39*(8), 2255–2276.

Kohtamäki, M., & Rajala, R. (2016). Theory and practice of value co-creation in B2B systems. *Industrial Marketing Management, 56*, 4–13.

Lieder, M., & Rashid, A. (2016). Towards circular economy implementation: A comprehensive review in context of manufacturing industry. *Journal of Cleaner Production, 115*, 36–51.
Liu, Z., & Stephens, V. (2019). Exploring innovation ecosystem from the perspective of sustainability: Towards a conceptual framework. *Journal of Open Innovation: Technology, Market, and Complexity, 5*(3), 48.
Lozano, R. (2018). Sustainable business models: Providing a more holistic perspective. *Business Strategy and the Environment, 27*(8), 1159–1166.
Lüdeke-Freund, F. (2010). Towards a conceptual framework of 'business models for sustainability'. In R. Wever, J. Quist, A. Tukker, J. Woudstra, F. Boons, & N. Beute (Eds.), *Knowledge Collaboration & Learning for Sustainable Innovation*, Delft, 25–29 October 2010.
Lusch, R. F., & Vargo, S. L. (2006). Service-dominant logic: Reactions, reflections and refinements. *Marketing Theory, 6*(3), 281–288.
MacArthur, E. (2013). Towards the circular economy. *Journal of Industrial Ecology, 2*(1), 23–44.
Moore, J. W. (2006). Animal ecosystem engineers in streams. *BioScience, 56*(3), 237–246.
Rockström, J. (2015). Bounding the planetary future: Why we need a great transition. *Great Transition Initiative, 9*, 1–13.
Romero, D., & Molina, A. (2011). Collaborative networked organisations and customer communities: Value co-creation and co-innovation in the networking era. *Production Planning & Control, 22*(5–6), 447–472.
Stead, J. G., & Stead, W. E. (2014). *Sustainable strategic management*. Routledge.
Strauss, A., & Corbin, J. (1994). *Grounded theory methodology: An overview.*
Stubbs, W., & Cocklin, C. (2008). Conceptualizing a "sustainability business model." *Organization & Environment, 21*(2), 103–127.
Vargo, S. L., & Lusch, R. F. (2004). The four service marketing myths: Remnants of a goods-based, manufacturing model. *Journal of Service Research, 6*(4), 324–335.
West, S., Müller-Csernetzky, P., & Huonder, M. (2018). Ecosystems innovation for service development. In *Practices and tools for servitization* (pp. 363–385). Springer.
Yin, R. K. (1994). Discovering the future of the case study. Method in evaluation research. *Evaluation Practice, 15*(3), 283–290.

Unleashing Innovation for a Sustainable World Session

Servitisation for Energy Transition and Circular Economy: Learnings from the CaaS and EaaS Business Cases

Dimitris Karamitsos, Alain Schilli, and Christophe Rynikiewicz

Abstract 35% of Green house Gases (GhG) savings to meet the Paris climate targets can be achieved through energy efficiency. Unfortunately, strong demand market barriers slow down market adaptation of energy efficient solutions. Servitisation, or the business model of paying for a service received instead of purchasing the asset, represents a strong solution to accelerate the implementation of higher energy efficient solutions, while also fostering a circular economy mindset and therefore reducing the economy's dependence and consumption of raw materials. Several companies across the globe have started implementing the model, and results are promising. Customers benefit from instantaneous energy consumption reductions while providers gain control to accelerate the implementation of innovative climate resilient technologies on the markets. Nonetheless, implementing servitisation in business operations is complex; it requires a shift in company revenue models, a much more active on the ground presence, digital solutions, reliable measurements, state of the art maintenance and repairs, and optimised supply chains, while also requiring for solution providers to develop the right financing mechanism to fund the implementation of the solutions and their operations. Implementing servitisation differs considerably from the classic sales business models. However, beyond these challenges, removing the upfront investment requirement and all perceived risks of operations from the customer's table is a highly competitive selling argument to convince clients to engage with climate resilient solutions.

Keywords Servitisation · Efficiency-as-a-service · Cooling-as-a-service

1 Introduction

According to the IEA, 35% of all GhG savings to meet climate targets can be achieved through Energy Efficiency (EE) alone. Unfortunately, strong market barriers slow down market adaptation of EE solutions; these can be summarized as (1) The higher

D. Karamitsos (✉) · A. Schilli · C. Rynikiewicz
BASE Stiftung, Elisabethenstr. 22, 4051 Basel, Switzerland
e-mail: eaas@energy-base.org; dimitris.karamitsos@energy-base.org

J. Meierhofer et al. (eds.), *Smart Services Summit*, Progress in IS,
https://doi.org/10.1007/978-3-031-36698-7_6

upfront investment of EE solutions, (2) EE solutions higher perceived risks, (3) Conflicting customers' investment priorities, (4) Uncertainties in maintenance and supply chain risks, (5) Complexities of technologies and compatibility of aggregated solutions and upgrades (Della Maggiora et al., 2022).

The dramatic increase in direct and indirect carbon emissions from the use of inefficient technologies across sectors and regions creates the urgent need to shift from power-hungry equipment to more efficient solutions that rely on cleaner technologies. This is all the truer in light of the latest IPCC report released in August 2021, along with the constant rise in electricity prices across Europe.

Time is of essence; there is an urgent need to reduce energy demand, cut pollution from industrial processes and ensure that efficient systems are affordable to all those who need them. Even more so, there is an imperative need to drastically change how businesses operate, to shift to a more circular mindset and reduce our use of raw materials.

Beyond technology innovation, business model innovation combined with digitalisation can be a strong enabler to accelerate the market transition to energy efficient solutions (Shaun et al., 2022). Furthermore, it permits embedding additional value to address the circular economy. Among these, servitisation, which is the shift from selling assets to selling output/performance contracts instead, can be a strong enabler which effectively tackles the demand barriers mentioned earlier.

The servitisation business model has the potential to align the interests of businesses, people and the planet, with all benefiting from social, economic and environmental gains. The maturity of the solutions and current implementation varies across sectors (Della Maggiora et al., 2022). Following the latter, this paper will analyze the below hypothesis:

(1) Does servitisation serve as a strong approach to tackle the market demand barriers to scale up energy efficient solutions and circular economy.
(2) Does the model perform economically and environmentally for large scale as well as smaller assets.

Within the research, the team will use a deductive case study methodology to yield the conclusions. Furthermore, within the analysis, the team will review what are the challenges in adopting the model, what are the aspects that may support in implementing and scaling it, and which sector could have a particularly large impact on EE improvement and climate impact (Fig. 1).

The following section outlines the core definition of the business model (Motmans et al., 2020).

2 The Theory of the Servitisation Business Model

Servitisation (or Servitization), is an innovative business model which transforms the traditionally product-focused business model driven by mass production and linear sales, into a service-focused model dedicated to the performance of the product,

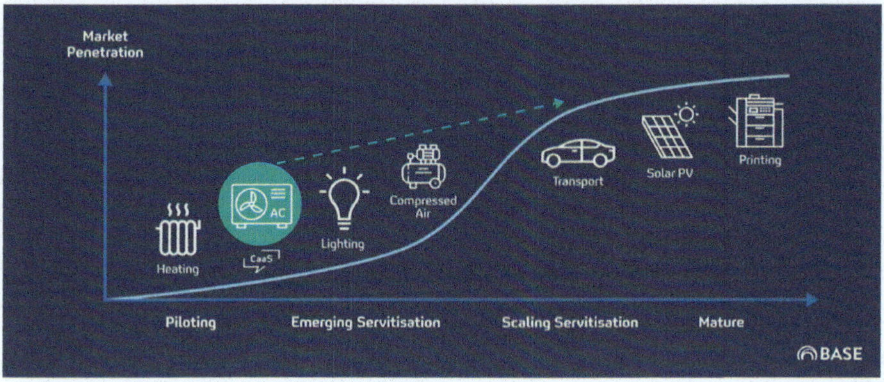

Fig. 1 Product as a service adaptation across various sectors

instead of the product itself, and promoting an extended lifecycle of the asset. Another way of framing it, is the shift from a transaction model to a subscription model, for which (especially for hardware equipment) an increase in performance, a reduction of raw material usage and critically recyclable materials drives higher margins.

With servitisation, upfront investment in deployed technologies is removed from the user, who instead pays per unit of service consumed (cooling, heating, lighting, compressed air, energy, etc.). Ownership of the asset stays with the solution provider who is responsible for implementing, operating and maintaining the asset. Performance risks remain with the solution provider who has the incentive to provide a reliable solution. With the right contracting, the model provides the right incentives for efficient and thoughtful consumption behaviors, which can be especially impactful when associated with clean and energy efficient solutions.

Through literature, servitisation holds various names; for specific products, the model is called "Product as a Service" (PaaS), while it is also called "Efficiency as a Service" (EaaS) when dedicated to energy efficient technologies, and "Cooling as a Service" (CaaS) when applied to the cooling sector. Within this paper, the authors will focus on EaaS and CaaS to study the impact of the model on accelerating the deployment of EE and reducing emissions while being economically viable.

Efficiency as a Service (EaaS) is a European H2020 funded initiative, coordinated by BASE, focusing on deploying the model of servitisation for energy efficient solutions in Belgium, Spain and the Netherlands. Cooling as a Service, an international initiative led by BASE as well, focused on designing and deploying the model of servitisation in cooling within emerging economies. Both models have now raised global interest.

3 Research and Deployment Methodology

To answer the questions of the research, the team focused on four pillars of activities.

(1) **Market analysis of country/geography**: determine the market barriers, the local policies, market demand per technology/industry, respective impact, local risks, possible subsidies, and key stakeholders' analysis to deploy energy efficient solutions.
(2) **Design, develop and test the tools needed to deploy servitisation in cooperation with key stakeholders**: as per feedback of the industry, this included the adequate financial structures, economic model aligned with IFRS15, standardized off-balance contract aligned with IFRS16 regulation, risk analysis tool, and customer documentation.
(3) **Capacity building and incubator**: engage and train key stakeholders to understand, implement and tailor the model to their needs. Within the CaaS initiative, this included an incubator to accelerate companies in deploying the model in their operations. Companies were selected within the incubator following a specific criterion, to analyse impact and stress test the model.
(4) **MRVs**: Share best practices in measuring, reporting and verifying the GHG savings from the solutions implemented, and contributions to net zero strategies, to yield conclusions on the climate impact.

With regards to MRVs, across regions and industries it is not yet common practice to measure the real performance of solutions implemented and track potential further improvements transparently. Servitisation brings the economic incentive to reliably measure the consumption of solutions implemented: generated revenues collected for the companies supplying the service depend on it.

4 Case Studies Selection CaaS–EaaS

4.1 Geographical Selection Criteria

Within the market analysis, the geographies for EaaS and CaaS were selected as per the following criteria: (1) Energy price of the region, (2) Climate strategies and targets of the region, (3) Economic and political stability of the region, (4) Market opportunity (real estate, industrial, private vs public sector), (5) Presence of solution providers on the ground.

These factors were selected to measure the impact of the model in specific regions and evaluate whether any additional measures are needed to further increase the value of the model (economically, socially, environmentally and from a governance perspective). For instance, within CaaS, countries selected included Argentina, Costa Rica, Grenada, India, Nigeria, Singapore, and South Africa. While for EaaS (being a European H2020 funded initiative), countries included Spain, Belgium and The

Netherlands due to the high price of electricity in 2018 and the large presence of SMEs which could benefit from servitisation.

4.2 Solutions Selection Criteria

The applications selected to design and deploy servitisation within EaaS and CaaS were based on the following criteria: (1) Strength of business case (payback time of solutions, investment requirement, location), (2) Capacity of implementations, (3) GHG emissions reduction potential, (4) Solutions risks (maturity of technologies, simplicity of implementation, currency risk, …), (5) Operational and maintenance costs of solutions, (6) Incentives or supporting policies.

EaaS selection applications—Based on the criteria reviewed, the following sectors and technologies were selected within EaaS (for Belgium, Spain, and The Netherlands) (Tables 1 and 2).

CaaS selection applications—Within CaaS, since the initiative focused on Cooling for emerging economies, BASE selected solutions from a range of different industries (and geographies) to stress test the model and evaluate the challenges for the various stakeholders in each application. The latter was combined with the countries mentioned in Sect. 4.1 and the criteria mentioned in Sect. 4.2. An incubator was launched to selected the applications which yielded: (1) Solar powered refrigerators and freezers (100–500 L capacity) in Nigeria, (2) Solar powered cold rooms (used in decentralised small-holder farmers areas and market place areas) in India and Nigeria, (3) HVAC systems in real estate (including offices, hotels, malls and hospitals) in Argentina/Costa Rica/Grenada/India/Singapore, (4) Refrigeration for industrial and commercial applications (frozen food companies, meat industry, large scale agriculture warehouses) in South Africa/India/Singapore.

Table 1 Sectors identified within EaaS (EaaS Project ResearchGate. Retrieved August 5, 2022). https://www.eaas-initiative.org/news/

	Belgium	Netherlands	Spain
Service	Food retail	Building related consumption	Hotels, hospitals
Industry	Food, chemical and plastic	Industry, fishing	Agri., food and beverages

Table 2 Identified priority technologies within EaaS (EaaS Project ResearchGate. Retrieved August 5, 2022). https://www.eaas-initiative.org/news/

	Belgium	Netherlands	Spain
Services	LED, cooling, solar	Heating, cooling, solar	Solar PV, LED, heating, cooling
Industry	LED, cooling, compress air	Heating, cooling, compress air	Industrial cooling, water pumps

5 Key Takeaways Contributions

Across applications, industries, and geographies, EaaS and CaaS case studies showed a strong potential to increase energy efficiency, combine asset operations with clean energy production and incentivizing providers to implement circular economy principles to extend their assets lifecycle and optimise returns on particular investments and unlock sustainable green financing. Due to the length of the paper, a high-level summary of these is listed in Table 3; the team can provide more information upon request. The key advantages and take aways observed from the market are listed in Table 4, while the reader is invited to read the case studies outline in detail on the CaaS-initiative website.

Table 3 Case studies and incubator high level summary

Application	Size	Country	High level observations
HVAC in real estate	400–1300 Tons of Refrigeration (TR)	Grenada/Costa Rica/Argentina/ India/Singapore	With large investments, credit risk is a bottleneck to finance projects in riskier economies. However, in growing and stable economies they are showing strong results and growth
Cold rooms in agricultures (adaptation projects)	5 Metric Tons (MT)	India/Nigeria	It is key to set up the rooms in areas which have a high useability rate, or rent the rooms to cooperatives that engage multiple small-holder farmers. In such cases the business model shows to be extremely competitive and bring strong value to the markets—reducing crop spoilage from 40 to 60% down to less than 10% (Motmans et al.: YourVCCA)
Fridges/ freezers	100–500 L	Nigeria	The model works well for small business owners that look to use multiple fridges. In the B2C market there is a strong ownership philosophy, and leasing model seems to be more accepted
HVAC in industry and refrigeration	400–1100 TR	South Africa/ India/Singapore	The model works particularly well. The providers are able to supply the solution combined with solar energy, which is a strong value adder when energy prices are high, the grid is unreliable and mainly coal powered; hence yielding strong emission reductions

Table 4 Key advantages of servitisation

Advantages	Why?
No CAPEX barrier for accessing solutions	Customers only pay for the use of a service rather than purchasing and owning the equipment. Across the world data shows that customers purchasing decisions are highly correlated to upfront investment requirements
Energy efficiency, and performance	As customers pay a fixed fee per service consumed, the more efficient a provider is, the more their respective profit margin and market competitiveness. Hence, the model promotes efficiency within existing contracts, and in cold rooms applications in agriculture, it also encourages more sustainable solutions via the use of thermal storage for cold rooms; and the reduction of batteries use
Increased usages of renewables	By shifting the asset ownership to the solution provider, it brings incentives for the providing company delivering the solution to use renewables (e.g., solar energy) in regions where the grid is unstable. This action reduces the dependency on fossil fuel grids, but also the use of diesel generators (for cold rooms), yielding a more sustainable economic model. Furthermore, where applicable it can generate additional profits by including carbon credits generation
Circular economy and optimised maintenance cycles	The model brings an economic incentive for companies to design components for long-term use, by optimising designs and deploying modular solutions which can be easily repaired or even recycled at the end of their life—with ideally local supply chains when possible—further increasing the profitability of the solution (ROI) while delivering a reliable solution to the end customer. Aligned with the latter, it encourages companies to design for optimised maintenance cycles to reduce the OPEX and CAPEX costs for delivering the solution and boosting profits while delivering state of the art service to the customer (Ellen McArthur Foundation, 2022). As such it incentivises companies to implement a circular economy approach and effectively tackles some of the barriers that slow down the adoption of Circular Economy (WeForum, 2022)
Optimised system sizes	The model incentivises the provider to size solutions fit for purpose which limits "over-designs". The latter always occurs in classic sales channels
Stranded assets	The model also drastically reduces the risk of stranded assets. This tends to happen when a user goes bankrupt or does not have the capital to repair a defective component or the network, or lack of adequate knowledge to do so
Useability rate	Depending on the application (for example cold rooms) EaaS/CaaS encourages the provider to position systems in strategic locations, ensuring a high useability-rate of the room (+70%) to yield an appealing return on investment for the asset owner. This engages providers to deliver a holistic business solution from production points to market areas for smallholder farmers

(continued)

Table 4 (continued)

Advantages	Why?
Misalignment of incentives	EaaS and CaaS showed to be effective in tackling misalignment incentives between tenants and real estate owner, which is typically a strong barrier in shifting to more efficient equipment in the build-in sector
Financing	By having a company manage multiple assets for diverse clients, data on the value of the business and its applications can be collected, which in turn increases financiers' trust to deploy funds for such applications, while eventually enabling for the asset and the contract to the client to act as a collateral of the financing provided. Research is at an early stage on assessing the financial aptitude of industrial firms to implement servitised earnings models (Kamp, 2020). However, financial structures can already be applied. Section 6 gives a more in-depth view of the market observations
Value chain	Value chains are more and more in the focus as a key driver for success. There are various geopolitical, consumer driven and legal developments offering business innovation opportunities to address customer's needs. Initially the focus was compliance risk in the supply chain as well as transparency and disclosure issues. Supplier code of conducts were addressing on a voluntary basis some of those issues. With NetZero pledges of the financial sector and upcoming legal frameworks with ESPR and CSRD (EU, 2022; ECOS, 2022) sustainable performance of manufacturing and sustainable products are key. This incentives resource-efficient and circular business models such as PaaS/EaaS/CaaS, reverse logistics, and on-demand production. Value chains are global for MNC as well as SME. Solid network and collaboration are essential for business success to secure flexibility and transparency. Servitisation business modeling is a collaborative approach, guided by co-creation. It offers the possibility to add value in the service development as well as in the supply chain
Digitalisation	The approach of selling performance to a customer unlocks the value in digitalization. It makes business sense for the provider to deploy the required digital solutions to accurately measure the performance of the asset, but also the degradation of these and when to repair, maintain or replace them. Digitalisation is a KEY component to also leverage financing for servitisation contracts, to proof-case their bankability. We recommend extending a research paper just on how to optimize and standardize digital measurements to fit the request of Financial Institutions to finance projects
Critical materials and chemicals	Entrepreneurs are more likely to implement systems in line with local regulations and trends in critical material management such as any chemicals used within solutions (such as refrigerant for cooling applications for instance)

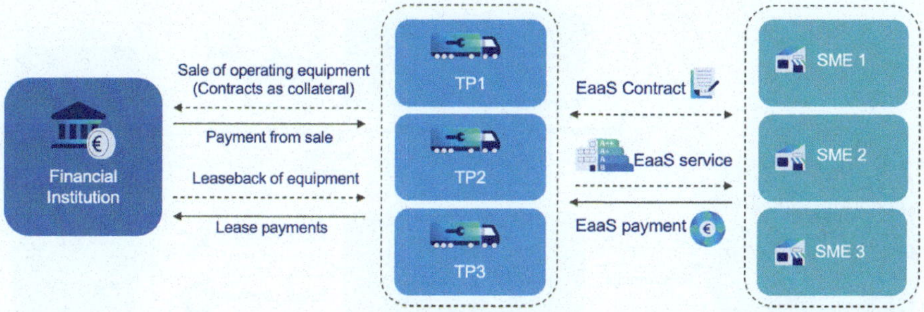

Fig. 2 EaaS sale and lease back financial structure (EaaS Project ResearchGate. Retrieved August 5, 2022)

6 Financial Structures

SME Energy efficiency projects might be small and can be complex to underwrite, and the transactional costs can be high compared to the size of the deals. A common approach is the "sale and lease back" mechanism (see Fig. 2), however it might be difficult and expensive to scale this financing method or to make it financially appealing to larger institutions (EaaS, 2020; Taya et al., 2021). Aggregation is envisaged as a strategy to group small individual projects together to make the task of evaluating the transaction and documenting the deals cost-effective. This facilitates more efficient financing deals with larger institutional investors. EaaS providers or SPVs (Special purpose vehicle) have the ability to aggregate energy efficiency projects to reach a scale where they are attractive for sale to large investors, for securitisation or access to competitive financing (Fig. 3). It remains, however, a challenge to find the financiers that invest in such opportunities, and more standardised rules and capacity building is required to bring them on board.

7 Challenges in Deploying Servitisation

The implementation of the different case studies enabled the study of the respective roles of actors along the servitisation journey as described by Aston Services Group at University of Birmingham (Baines et al., 2020). It was observed that companies are at various stages of maturity of understanding and deploying the approach.

The challenges which were observed in deploying servitisation revolve around a lack of knowledge on servitisation contracts (IFRS16), pricing models (IFRS15), financing solutions, along with operational changes which are required to deploy the approach. In some applications, customers also have a strong dependency to own equipment (with limited knowledge on the TCO of owning assets), and substantial capacity building needs to be done to expose the value of PaaS/EaaS/CaaS transparently (Bergman & Linder, 2022).

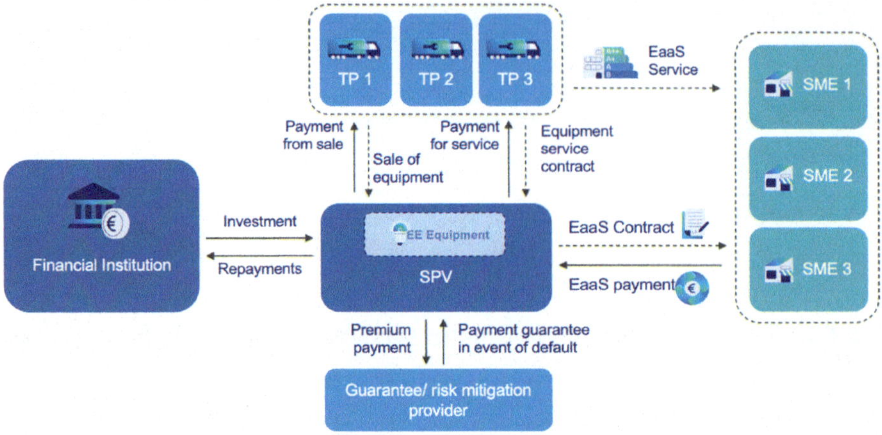

Fig. 3 EaaS SPV financial structure (EaaS Project ResearchGate. Retrieved August 5, 2022)

For servitisation to be successful, every stakeholder in the process needs to be engaged; with the right team on the ground, an innovative customer and a green financier seeking to deploy funds on decarbonising opportunities. Fortunately, as expertise to integrate the model is spreading across the markets, a growing number of companies are embracing the opportunity to integrate the approach, governments have a strong role to play, as users and as well to build capacity. As knowledge on the model expands, new industries are raising interest as well (batteries as a service for e-mobility, hybrid propulsions as a service on vessels, solar panels on bulk carriers to name a few).

8 Conclusion

The paper presents key findings from projects implemented in the building sector (real estate, commercial, industrial, data centers, …). There is no "one size fits all" method to implement CaaS or EaaS projects. It often takes time and iterations for the operations of the solution provider, its products and how their clients are approached. Common factors often apply: cf CaaS white paper.

However, it was observed that Servitisation can be a strong business model to accelerate cities' path to Net Zero, implement a circular economy, as well as speed up the implementation of adaptation projects (in agriculture). The model showed to perform economically and environmentally for large scale as well as smaller assets. Nonetheless, more efforts must be brought forward to engage the financial institutions in financing these projects. Digitalisation can be one leverage to onboard more financing, by using data to showcase the bankability of such projects and their impact on climate. Several initiative exist to support stakeholders adopt the business model; the global Servitisation for Energy Transition (SET) Alliance, provides tools and experts' know-how for companies to shift to servitisation, and the WEF is also launching a Toolbox of solutions at the COP27, which among other solutions also includes servitisation to accelerate the transition to Net Zero. In this mission to achieve a more climate resilient future, the role of every entity is key; partnerships, collaboration and a change in classic business models is key to achieve success.

References

Baines, T., Bigdeli, A., Sousa, R., & Schroeder, A. (2020). Framing the servitization transformation process: A model to understand and facilitate the servitization journey. *International Journal of Production Economics, 221.*

Bergman, E., & Linder, M. (2022). *Product as a service in the circular economy.*

Della Maggiora, C., Karamitsos, D., & Motmans, T. (2022, February). *Servitisation of the cooling industry: CaaS white paper.* Retrieved August 22, 2022, from https://www.caas-initiative.org/wp-content/uploads/2022/05/Cooling-as-a-Service-White-Paper-1.pdf

EaaS. Retrieved August 6, 2022, from http://eaas-initiative.org

EaaS Project ResearchGate. Retrieved August 5, 2022, from https://www.researchgate.net/project/eEaaS-Efficient-Equipment-as-a-service

Ecodesign for intermediate products: ECOS position paper. Retrieved September 29, 2022, from https://ecostandard.org/wp-content/uploads/2022/07/ECOS-Position-Paper-ESPR-for-intermediate-products-July-2022.pdf

Ellen McArthur Foundation: Kaer CaaS Business Case and Circular Economy. Retrieved September 26, 2022, from https://ellenmacarthurfoundation.org/circular-examples/cooling-as-a-service-kaer

EU Commission. *Ecodesign for sustainable products regulation: ESPR proposal.* Retrieved September 29, 2022, from https://environment.ec.europa.eu/publications/proposal-ecodesign-sustainable-products-regulation_en

Kamp, B. (2020). Assessing the financial aptitude of industrial firms to implement servitised earnings models. *International Journal of Business Environment, 11*(1), 2020.

Motmans, T., et al. (2020). Cooling as a service: Unlocking demand for clean and efficient cooling. In *ECEEE Industrial Summer Study.* Retrieved August 15, 2022, from https://www.eceee.org/library/conference_proceedings/eceee_Industrial_Summer_Study/2020/5-business-models-finance-and-investment/cooling-as-a-service-unlocking-demand-for-clean-and-efficient-cooling/

Motmans, T., et al. Retrieved October 22, 2022, from https://yourvcca.org/

Shaun, W., Jürg, M., & Utpal, M. (2022). *Smart services summit.*

Taya, M., et al. (2021, May). Servitisation model in energy efficient equipment: Implementation study in the Netherlands, Belgium and Spain. In *Proceedings of the Spring Servitisation Conference.*

WeForum. *Circular economy for net-zero industry transition, "the barriers to circular approaches".* Retrieved August 22, 2022, from https://www.weforum.org/circular-economy-for-net-zero/barriers

Assessment of the Influence of Grid Resolution on CO_2 Reduction in Route Optimization Services Using Reinforcement Learning

Mohammad Hossein Moradi, Martin Brutsche, Markus Wenig, Uwe Wagner, and Thomas Koch

Abstract Following the Paris Climate Agreement, the maritime industry has committed to reducing its GHG emissions by 50% by 2050 (compared to 2008). In this sense, the present work pursues this goal by focusing on an improved route optimization method using Reinforcement Learning (RL). A detailed comparison between RL and the conventional approach is carried out in this study. Besides RL, Dynamic Programming (DP) is also used to establish the benchmark. The influence of different grid resolutions and dynamic weather on effective CO_2 reduction is analyzed. It is observed that these two aspects can play a significant role in the route optimization results. Furthermore, the results show that RL as a model-free approach offers a great advantage for these considerations.

Keywords Marine route · Optimization · CO_2 reduction · Reinforcement learning · Artificial intelligence · Grid resolution

M. H. Moradi (✉) · U. Wagner · T. Koch
Institute of Internal Combustion Engines, Karlsruhe Institute of Technology, Rintheimer Querallee 2, Geb. 70.03, 76131 Karlsruhe, Germany
e-mail: mohammad.moradi@kit.edu

U. Wagner
e-mail: uwe.wagner@kit.edu

T. Koch
e-mail: thomas.a.koch@kit.edu

M. Brutsche · M. Wenig
Winterthur Gas & Diesel Ltd., Winterthur, Switzerland
e-mail: martin.brutsche@wingd.com

M. Wenig
e-mail: markus.wenig@wingd.com

J. Meierhofer et al. (eds.), *Smart Services Summit*, Progress in IS,
https://doi.org/10.1007/978-3-031-36698-7_7

1 Introduction

The environmental crisis encourages all sectors of industry to think of a solution to reduce CO_2. Following the Paris Climate Agreement, the Maritime industry is committed to reducing its GHG emissions until 2050 by 50% (compare to 2008) (Mepc, 2018). In the mid-term future, due to the anticipated scarcity of carbon-neutral or zero-carbon fuels, this requires an overall optimization of all involved systems. Several aspects make CO_2 reduction possible. In addition to increasing engine and hull efficiency, the condition of the ship's operation can be improved and optimized. Since the fuel consumption in ship operation is quite dependent on weather conditions, optimizing the route of the ship can lead to a significant saving in fuel consumption and subsequently reduce CO_2 emission (Zaccone et al., 2018).

Route optimization for the ships has been carried out by various studies. Route optimization in maritime shipping is mainly due to avoiding rough sea conditions and consequently saving fuel consumption. In general, these studies can be divided into two different categories depending on the consideration of weather data. In the studies that took weather data into account, in addition to minimizing distances also the influence of weather conditions on fuel consumption is analyzed. However, all methods have various advantages and disadvantages and may differ in aspects such as calculation method or the possible level of detail. Zis et al. (2020) have compiled a detailed list of the various studies in this field. Various methods have been used for route optimization, including Isochrone methods (Hagiwara, 1989; Lin et al., 2013), dynamic programming (Calvert, 1990; Chen, 1978; Dijkstra, 1959; Wang et al., 2019; Zaccone et al., 2018), and evolutionary algorithms (Hinnenthal & Günther, 2010; Lee et al., 2018; Maki et al., 2011; Marie & Courteille, 2009; Vettor & Guedes Soares, 2016; Zaccone et al., 2016), etc. Other methods which can be applied for route optimization are Machine Learning methods (Beşikçi et al., 2016; Coraddu et al., 2017; Du et al., 2019; Gkerekos & Lazakis, 2020; Perera & Mo, 2016; Zheng et al., 2019). Machine learning can be divided into subclasses depending on the type of learning method and training data. The most important representatives have Supervised Learning (SL), Unsupervised Learning (UL), and the method used in this work, namely Reinforcement Learning (RL). RL is not SL because there is no direct labeling of the data and it is also not UL because there is a signal, namely reward, which is missing in unsupervised learning (Sutton & Barto, 2018). Moradi et al. (2022) have proposed a new method for route optimization. They have used RL as an optimization method in their study. The advantages of RL over conventional methods have been explained in detail in this study. These advantages can be summarized in two important points.

1. RL as a model-free method can work with fine grids.
2. RL allows the consideration of dynamic weather data.

The characteristics of RL methods enable these advantages. In Moradi et al. (2022), these advantages were mentioned, but there was no detailed analysis of the influences of these factors on the accuracy of the results. In this respect, the present study analyses these factors in more detail and compares DP (the main conventional method) and RL.

The importance of Digital Twins (DT) in decision-making processes has been demonstrated in previous studies (Galeno et al., 2022; Schweiger et al., 2022; West et al., 2021). In this way, this study presents another aspect of DT for decision-making to reduce fuel consumption and increase sustainability.

2 Methodology

The current study is concerned with analyzing the influences of a finer grid system and dynamic weather data on the results of route optimization for the vessels. Figure 1 shows the differences between the two grid resolutions. The weather data are retrieved only for the grid point coordinates, and other coordinates take the same weather data as the grid point standing afterward. In other words, the fewer steps there are for the grid, the fuel consumption for the greater distance calculated with the same weather condition, which increases the potential for incorrect calculations. Static and dynamic weather data differ in the time by which the weather data is retrieved. In static weather data (independent of previous steps) an estimated time for each stage is considered. In dynamic consideration, depending on the vessel speed and navigated route, the actual arrival time for each grid point is calculated and the weather data for this time is retrieved from the API. More detailed explanations are given in Moradi et al. (2022).

DP is used as a conventional method and the results are compared with the results of RL. In DP and RL, the Markov-Decision process is applied, with the difference that DP is a model-based method and RL is a model-free method, i.e. DP needs holistic information about the environment and requires very high computing capacity. Furthermore, the entire state space must be stored in the memory, which grows exponentially with each additional size or finer discretization. RL on the other hand is a model-free method, i.e. the RL agent does not have any information about

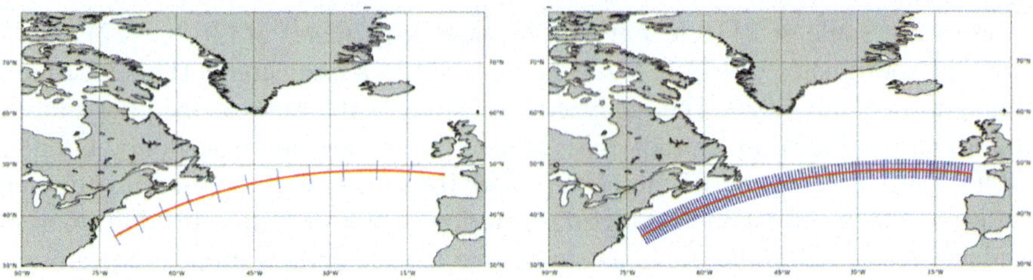

(a) (b)

Fig. 1 The structure of the grid in two different resolutions; **a** rough grid and **b** fine grid

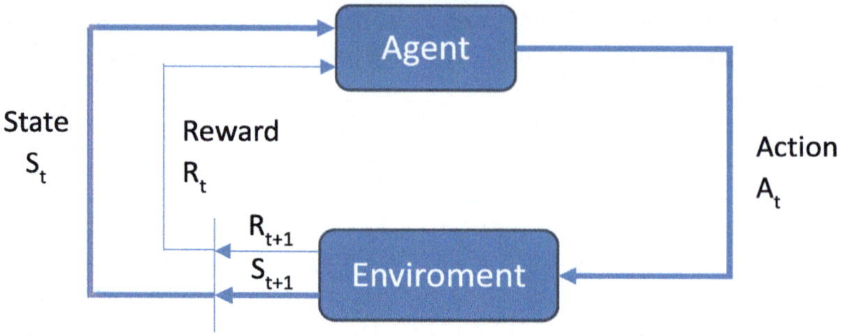

Fig. 2 The schematic of RL method (Sutton & Barto, 2018)

the environment at the beginning. In the course of training, it explores the environment, gathers the experience from the environment through rewards, and uses the gained experience for new actions. The scheme of RL algorithms is shown in Fig. 2.

To construct the environment of the RL algorithms as well as the probability matrices of DP-Method, a two-stage neural network (as in Moradi et al., 2022) is used. A detailed description of the Artificial Neural Network (ANN)-models as well as a detailed explanation of the differences between DP and RL can be found in Moradi et al. (2022). The approach of this study shows the significant potential of Digital twin in CO_2 reduction in the shipping industry. For this purpose, a simulation is developed based on the monitoring data with artificial neural networks and then this model is applied for route optimization purposes. Without suitable monitoring of the real data, the development of the model and consequently the optimization of the route was not possible. This application highlights the importance of monitoring data for future calculations, where machine learning must be used to create a model that is both fast and accurate. Figure 3 shows the workflow overview of this study.

3 Result and Discussion

A case study route of 5346.17 km with the start coordinates of (36.1, −72) and the endpoint coordinates of (48.1, −7.2) is analyzed as a case study. The start time is midnight on 5 July 2020. The analyzed case study is the same as the case study in Moradi et al. (2022). In Moradi et al. (2022), Deep Deterministic Policy Gradient (DDPG) was used as the RL algorithm, and 106 stages were applied for the optimization, which corresponds to a grid resolution of almost 50 km. In addition to the available results from the previous study, two further optimizations are performed in this paper, namely, with DDPG (RL) and only 11 stages (a resolution of almost 500 km) and with DP and 11 stages. The number of states for 11 stages and without variation of speed is 112, whereas 1157 states are needed for 106 stages. As mentioned before, DP is a model-based method and is typically applied to the lower number of stages. Furthermore, due to its nature, DP is only capable of handling static weather

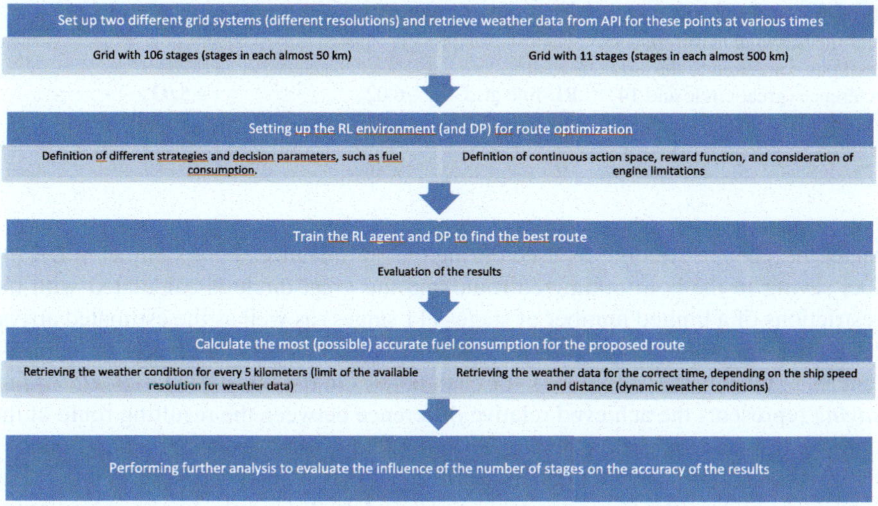

Fig. 3 Workflow overview

data, i.e., the results of RL may differ from the results of DP with the same number of stages (11 stages) since RL can consider dynamic weather data. The consideration of dynamic weather data is not possible with RL, as an extremely high number of states is required for such analysis. After optimizing the route with these three methods, is analyzed if the calculation of the fuel consumption can be faulty due to a rough grid and static weather data. Analyses are carried out for the scenario without a time limit, which is the reason that the minimum speed is chosen in all the optimization methods as the best feasible speed, which is almost 14 knots. The structure of the grid and the proposed route of the different methods are presented in Figs. 1 and 4, respectively.

Table 1 shows the predicted and calculated fuel savings in different methods compared to the reference (great circle and 14 knots). The negative values show a reduction in fuel consumption compared to the reference. *Predicted fuel saving* is the predicted relative difference between the optimized route and the great circle

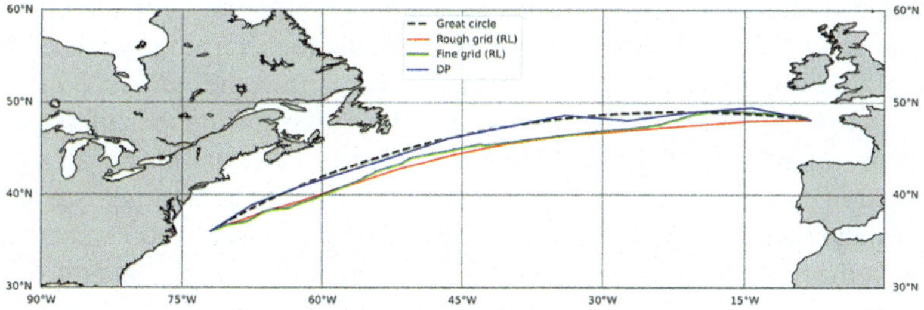

Fig. 4 The proposed routes from different optimizations conditions

Table 1 The calculated and real fuel savings in different methods compared to the reference (great circle and 14 knots)

Method	Predicted fuel saving (%)	Calculated fuel saving (%)
RL-fine grid	−6.02	−5.93
RL-rough grid	−5.78	−3.81
DP	−8.92	−0.50

under the same restrictions as used during the optimization. For example, in DP, the fuel saving in the both optimized route and the great circle is calculated with the restrictions of a limited number of stages (11 stages) as well as the estimated arrival time for each point (instead of the actual arrival time), which results in inconsistencies in the weather conditions for each point. On the other hand, *calculated fuel saving* represents the achieved relative difference between the resulting route of the optimization and the great circle, evaluated with the most accurate data as possible. For this calculation, weather data is retrieved along the proposed route and the great circle every 5 km (the finest available weather data ("stormglass.io,"), compared to 500 km used for optimization with DP or respectively 50 km for RL). Furthermore, depending on the ship's speed and the distance between two points the weather data for the actual arrival time is retrieved, which enables the calculation of the most accurate value of fuel consumption. This Table shows that although the optimization with DP leads to a saving of almost 9%, in reality, it is only 0.5%. This shows the significant influence of the rough grid and the static weather data, which lead to an error of almost 8.5% in the calculations. However, the results of RL with a coarse grid (same number of steps as DP) show that optimization gives only 5.78% (about 3% less than DP) savings in fuel consumption, but the actual fuel saving is 3.81% (about 3% more than DP). The better result is due to the consideration of dynamic weather data which is possible with RL. When the optimization is performed with RL and finer grid resolution, almost identical values (almost 6%) are obtained in the simulation and reality. These results show the indispensable advantages of RL over conventional methods. In other words, the simplifications in other methods make the results of the optimizations unreliable. Consequently, optimization can even lead to a deterioration in fuel consumption, which means there is no guarantee that fuel saving is occurring.

As a further analysis, the fuel consumption along the great circle is calculated at different numbers of stages. In this sense, the great circle is divided into 5 km resolution between the start and endpoints. Then the fuel consumption is calculated with a different number of stages. This analysis clarifies the influence of the grid fineness on the accuracy of the results to find a sufficient number of stages for a reasonable calculation of fuel consumption. All these calculations are based on a ship speed of 14 knots. It should be mentioned that dynamic weather is considered in this analysis. Figure 5 shows the results of this analysis. The calculated error represents the relative error between the different number of stages and the reference (calculation of fuel consumption with weather data for the finest resolution, i.e., every five kilometers). On the left side, the results of up to 1000 stages are shown, on the

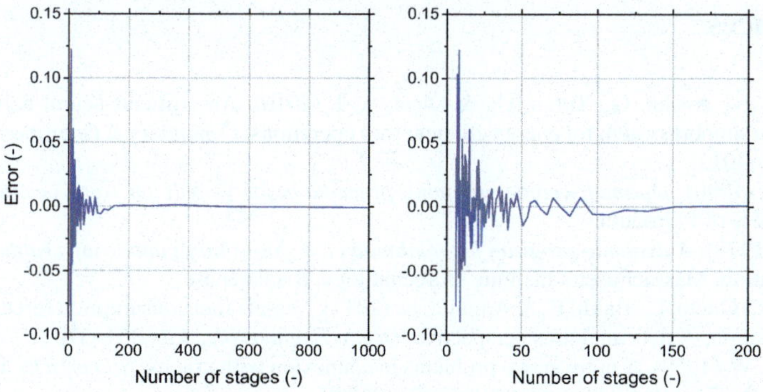

Fig. 5 The error in the calculation of fuel consumption in various number of stages

right side the same diagram is enlarged, and only the first 200 stages are shown. The results show that from about 100 stages onward there are almost no significant errors. Up to 50 stages, on the other hand, there is a large variation in the calculated fuel consumption. These results show again that the conventional methods can provide unreliable results due to the high computational effort and the limited number of stages.

4 Conclusion

In this paper, an extensive analysis of the influences of grid fineness and dynamic weather data on route optimization is performed. A benchmark of conventional and new methods is established. Dynamic Programming (DP) represents the conventional methods in this work and its results are compared with Reinforcement Learning (RL) results. There is an 8.5% discrepancy between the expected fuel consumption and the actual fuel consumption in DP, resulting in a fuel saving of only 0.5% by optimization. This error is due to the limited number of stages and the static consideration of weather conditions. RL, on the other hand, can achieve fuel savings of 3.81% with the same number of stages as DP, and up to 6% fuel savings in reality with a finer grid. The advantages of RL over the conventional methods, i.e., the possibility of considering the dynamic weather data as well as a finer grid system (due to the model-free approach), allow these improvements. The results of this study have shown that RL has significant potential for application in the route optimization of the vessel. With the application of monitoring data and by developing machine learning models based on this data, a method for analyzing different methods for route optimization of the ship has been developed.

References

Beşikçi, E. B., Arslan, O., Turan, O., & Ölçer, A. I. (2016). An artificial neural network based decision support system for energy efficient ship operations. *Computers & Operations Research, 66*, 393–401.

Calvert, S. (1990). *Optimal weather routeing procedures for vessels on trans-oceanic voyages*. University of Plymouth.

Chen, H. (1978). *A dynamic program for minimum cost ship routing under uncertainty*. Doctoral dissertation, Massachusetts Institute of Technology, Ris database.

Coraddu, A., Oneto, L., Baldi, F., & Anguita, D. (2017). Vessels fuel consumption forecast and trim optimisation: A data analytics perspective. *Ocean Engineering, 130*, 351–370.

Dijkstra, E. W. (1959). A note on two problems in connexion with graphs. *Numerische Mathematik, 1*(1), 269–271. https://doi.org/10.1007/bf01386390

Du, Y., Meng, Q., Wang, S., & Kuang, H. (2019). Two-phase optimal solutions for ship speed and trim optimization over a voyage using voyage report data. *Transportation Research Part b: Methodological, 122*, 88–114.

Galeno, G., Barth, L., Ehrat, M., & Demiriz, U. (2022). Value creation with digital twins in the end-of-lifecycle phase of smart products: Applied data resources in academic literature. In *Smart services summit* (pp. 85–94). Springer.

Gkerekos, C., & Lazakis, I. (2020). A novel, data-driven heuristic framework for vessel weather routing. *Ocean Engineering, 197*, 106887.

Hagiwara, H. (1989). *Weather routing of (sail-assisted) motor vessels*. Ph.D. thesis, Technical University of Delft. https://ci.nii.ac.jp/naid/20000749977/

Hinnenthal, J., & Günther, C. (2010). Robust Pareto-optimum routing of ships utilising deterministic and ensemble weather forecasts. *Ships and Offshore Structures, 5*(2), 105–114. https://doi.org/10.1080/17445300903210988

Lee, S.-M., Roh, M.-I., Kim, K.-S., Jung, H., & Park, J. J. (2018). Method for a simultaneous determination of the path and the speed for ship route planning problems. *Ocean Engineering, 157*(09), 301–312. https://doi.org/10.1016/j.oceaneng.2018.03.068

Lin, Y.-H., Fang, M.-C., & Yeung, R. W. (2013). The optimization of ship weather-routing algorithm based on the composite influence of multi-dynamic elements. *Applied Ocean Research, 43*, 184–194. https://doi.org/10.1016/j.apor.2013.07.010

Maki, A., Akimoto, Y., Nagata, Y., Kobayashi, S., Kobayashi, E., Shiotani, S., … Umeda, N. (2011). A new weather-routing system that accounts for ship stability based on a real-coded genetic algorithm. *Journal of Marine Science and Technology, 16*(3), 311–322. https://doi.org/10.1007/s00773-011-0128-z

Marie, S., & Courteille, E. (2009). *Marine navigation and safety of sea transportation*. CRC Press/Balkema.

Mepc, I. R. (2018). *304 initial IMO strategy on reduction of GHG emissions from ships*. London, UK: MEPC.

Moradi, M. H., Brutsche, M., Wenig, M., Wagner, U., & Koch, T. (2022). Marine route optimization using reinforcement learning approach to reduce fuel consumption and consequently minimize CO_2 emissions. *Ocean Engineering, 259*. https://doi.org/10.1016/j.oceaneng.2022.111882

Perera, L. P., & Mo, B. (2016). Data analysis on marine engine operating regions in relation to ship navigation. *Ocean Engineering, 128*, 163–172.

Schweiger, L., Meierhofer, J., Barbieri, C., & Rapaccini, M. (2022). End-to-end digital twin approach for near-real-time decision support services. In *Smart services summit* (pp. 67–75). Springer.

stormglass.io. https://stormglass.io/

Sutton, R. S., & Barto, A. (2018). *Reinforcement learning: An introduction* (2nd ed.). Cambridge, MA; London: The MIT Press.

Vettor, R., & Guedes Soares, C. (2016). Development of a ship weather routing system. *Ocean Engineering, 123*(3), 1–14. https://doi.org/10.1016/j.oceaneng.2016.06.035

Wang, H., Mao, W., & Eriksson, L. (2019). A three-dimensional Dijkstra's algorithm for multi-objective ship voyage optimization. *Ocean Engineering, 186*(2), 106131. https://doi.org/10.1016/j.oceaneng.2019.106131

West, S., Stoll, O., Meierhofer, J., & Züst, S. (2021). Digital twin providing new opportunities for value co-creation through supporting decision-making. *Applied Sciences, 11*(9), 3750.

Zaccone, R., Figari, M., Altosole, M., & Ottaviani, E. (2016). *Fuel saving-oriented 3D dynamic programming for weather routing applications.*

Zaccone, R., Ottaviani, E., Figari, M., & Altosole, M. (2018). Ship voyage optimization for safe and energy-efficient navigation: A dynamic programming approach. *Ocean Engineering, 153*, 215–224. https://doi.org/10.1016/j.oceaneng.2018.01.100

Zheng, J., Zhang, H., Yin, L., Liang, Y., Wang, B., Li, Z., … Zhang, Y. (2019). A voyage with minimal fuel consumption for cruise ships. *Journal of Cleaner Production, 215*, 144–153.

Zis, T. P., Psaraftis, H. N., & Ding, L. (2020). Ship weather routing: A taxonomy and survey. *Ocean Engineering, 213*, 107697.

Business Models of an AI Marketplace for Energy Systems with Focus on Demand Response

Braulio Barahona, Benjamin Bowler, Colin Gounden, and Antonios Papaemmanouil

Abstract Three marketplace archetypes of increasing complexity are conceptualised with business models for services that support the implementation of demand response programs.

Keywords Smart services canvas · Product-service systems

1 Introduction

The need to transmit data, information, and control actions across domains while protecting people's privacy, and the integrity of components, calls for novel tools and business models. Hussain et al. (2014) identified the lack of interoperability between business layers and the cyber-physical electricity grid and proposed that IT service providers establish service level agreements (e.g., service level and performance indicators to be monitored) as a means to increase coordination and trustworthiness. More recently, Kern et al. (2022) also highlighted the need for new business models driven by digitalisation technologies like AI. They described a prosumer platform and explained how AI could contribute within this platform in the form of tools for forecasting, for customer service (e.g., via a chatbot), and for intelligent monitoring and control of consumption. A more radical view on novel digital services is that of an *AI marketplace*, where marketplace users can access data consolidation workflows, securely share data across stakeholders, and ultimately deploy analytics workflows including state-of-the-art AI models. The term *AI marketplace* is not as established as those of *data space*, *data as a service*, *software as a service*, or *(application)*

B. Barahona (✉) · B. Bowler · A. Papaemmanouil
Lucerne University of Applied Sciences and Arts, Horw 6048, Lucerne, Switzerland
e-mail: braulio.barahonagarzon@hslu.ch

C. Gounden
Via Science AG, Suurstoffi 18B, 6343 Rotkreuz, Switzerland

J. Meierhofer et al. (eds.), *Smart Services Summit*, Progress in IS,
https://doi.org/10.1007/978-3-031-36698-7_8

platform as a service. It emerges as an ecosystem of digital products and services such as data markets where users can trade datasets or provide labelling and curation services. Portals where models can be downloaded or run in the cloud are also part of this trend. Given that power grids are critical infrastructure, they are subject to special regulations in terms of cybersecurity and data protection. Therefore, Berkhout et al. (2022) suggested the use of open, standardised architectures implemented on sovereign IT infrastructure as an alternative to data lake solutions of big tech firms. Several research projects funded by the European Union are developing prototypes with the vision to deliver demonstrations that drive acceptance of these ecosystems,[1] relevant works include those by Prado et al. (2020), Mehri and Tutschku (2017), Mehri (2019), and Ouyang et al. (2022) that address several technical aspects but do not focus on the business models. Aspects such as development and operation costs, data protection regulations, as well as data and information systems interoperability pose challenges that may be well addressed by AI marketplaces as described by Kumar et al. (2021), where a healthcare use case was described. Here, we focus on the development of digital services for energy systems delivered via AI marketplaces. In Sect. 2 we describe our approach with focus on the use of business model canvases. The resulting concepts for digital services are described in Sect. 3, challenges and next steps are discussed in Sect. 4.

2 Methods

Our approach consist of (1) preliminary selection of use cases; (2) mapping the value proposition to the system value chain; (3) analysis and design of digital services. The preliminary selection of cases within energy systems is based on heuristics relating to the need for digitalisation, the market needs, and the potential impact on decarbonisation. Then, we apply a system value framework specific to energy systems to map value exchanges. In the final step we explore business strategy tools, such as service dominant logic, and the value proposition canvas. Particularly, we focus on the use of value creation and smart services canvas to derive business models and service-specific characteristics.

Business intelligence analyses by Singh et al. (2022) pointed out multiple **X**-as-a-Service Business Models: they identified "demand response marketplace", "demand response", and "digital marketplace to connect producers and consumers" as central value propositions for **Flexibility-**, **Energy-**, and **Trading-**as-a-Service. Such servitization has been happening in other sectors, as it is seen as a means to create new revenue streams on the basis of digital technology. However, simply creating digital services is not a guarantee for successful innovation. Naturally, a first step in the service design is to understand the value exchange relations amongst the actors in a product-service ecosystem. On the basis of system maps, Verdugo Cedeño et al.

[1] A list of marketplaces with descriptions of their application domains, main attributes, actors, and current state can be found at www.gitlab.com/hslu_deep/ai-market-places.

(2019) applied knowledge discovery to guide the definition of requirements throughout the life-cycle of a service. In this work, product life-cycle is out of scope and we address the prototyping of the service using the electricity system value framework (Crofton et al. 2015) to formalise our domain knowledge. We then adopt service dominant logic design paradigms on the basis of West et al. (2018), where IoT-based service systems in manufacturing industries where analysed, indicating positive characteristics of the value co-creation, the service ecosystem, and the platform. Namely, in terms of value co-creation, we aim at integrating multiple resources and at facilitating multi-actor interactions. The service ecosystem should have a secured architecture and allow multiple roles. The service platform should be modular, allow integration of third party services, and have clear rules of exchange. Our final step consists of using a type of business model canvas, the Smart Service Canvas (SSC), described by Poeppelbuss and Durst (2019). It builds on value proposition canvas and proposes to model service concepts focusing on four dimensions, each with multiple parameters: **Customer Perspective** (Gains, Pains, Customer Routines and Jobs, Context of Customer Routines and Jobs, and Contextual Things and Data), **Value Perspective** (Create Value, Solve Problems, Smart Service, Analytical Capability, and Data), **Ecosystem Perspective** (Technical infrastructure and Digital Platform), and **Fit**(Revenue Model, Interaction Level, Smart Product). The developed archetypes for AI marketplaces and services designed are described in Sect. 3.

3 Results

System value from load decomposition to demand response (DR). Final energy consumers participating in DR programs seek to reduce their energy cost and footprint on the environment. DSOs seek to deliver electricity while complying with regulations that are evolving to promote the integration of higher shares of renewable energy into the grid. OEMs may want to expand their product-service systems. For example, manufacturers of digital electricity meters may develop cloud solutions to help their customers to manage and visualise their data. ICT companies come in as partners or technology providers. A digital marketplace can create new value propositions as illustrated in Fig. 1 where added value can come from model and/or data providers. Components of the DR application include: (*I*) Non-intrusive load monitoring (NILM) to decompose measured electricity consumption and identify appliances being used behind a metering point over a given sampling period to quantify their individual consumption. NILM models rely on large, labelled datasets that represent a diverse range of customer behaviours and appliance types; (*II*) Load curve forecasting supports the prediction of future network states. It benefits from analysis of a consumers historical data, for example, by aggregating forecasts of peak loads, allowing a system operator to predict when periods of congestion may occur, and hence curtail load or encourage users to shift their consumption to periods of lower system stress; and (*III*) Interaction between DSO and consumer for the utilisation

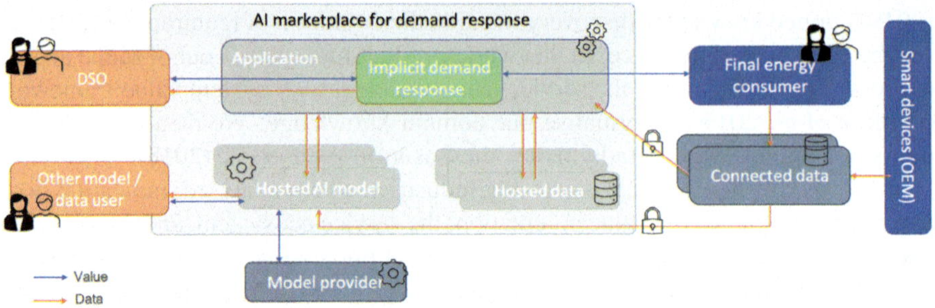

Fig. 1 Illustrative value and data flows in an AI marketplace for demand response (DR)

of data, creating of models, and adaption of the user's energy consumption patterns responding to signals from the DSO.

Archetype I: Static AI marketplace. This basic form of marketplace features a repository for models and data, and it provides a meeting place where users can offer or access these digital assets. Model descriptions can be stored in and downloaded from the marketplace. Data can be also be accessed for training or evaluating models. From the customer perspective, the principal role is the model user or the data user. From the value perspective, it is the model provider or the data provider; and from the ecosystem perspective, it is the market operator and the digital platform provider. Customers access the market because they would like to apply an AI model to a forecasting, detection, or classification problem. The customer may need to use the market because they lack the expertise to develop a reliable model in-house, or they may wish to use a model for a task for which mature models already exist, having been developed by others. By accessing an existing model, they may be able to reduce the development costs, or be able to access the most performant model for their task, therefore improving the probability of success of their solution. The customer may also access the market to download bigger datasets or more representative than those already in their possession.

The products offered are therefore: (1) trained model or a dataset for download, (2) hosting mechanism for storing models and data, and (3) a transaction between a model and/or data provider and the model and/or data user. The smart services offered by the market in relation to the smart products are: (1) operation of a digital platform that curates data and models; (2) execution of a trading process that allows a transaction to take place between model providers and model users; (3) value transfer between users of the market. Value is created through the application of trained models or data in the customer's operations. Value capture may also take place through cost avoidance as the customer is able to reduce effort in model training or take advantage of established data pre-processing workflows.

Additional roles and smart services that enhance the value captured by the market operator are envisaged in the static marketplace. For example, the market operator may independently validate the accuracy or performance of models or data, or may provide additional resources relating to them. These may include technical and sci-

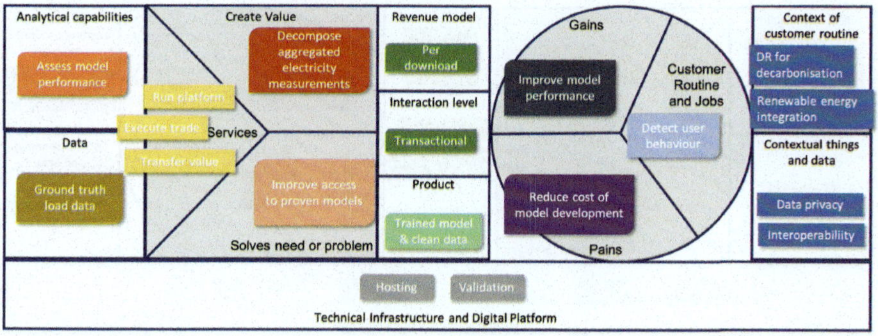

Fig. 2 Example static AI marketplace for non-intrusive load monitoring model and labelled data

entific content in the form of tutorials to guide users in the application of the models. Models for NILM are suitable for the static AI marketplace. The SSC in Fig. 2 illustrates the NILM application.

Archetype II: Dynamic AI marketplace. This archetype requires additional features that allow the marketplace user to: (1) run or retrain a model online within the digital platform; (2) connect their own external datasets to the marketplace so that they can be included in model retraining; (3) access and connect to external third-party datasets through the market so that the they can be included by the user in model retraining; (4) obtain insights into metrics relating to training, for example carbon impact of retraining models. Services and technical infrastructure offered by the digital platform provider include: (1) compute resources for model retraining; (2) interfaces for connecting external datasets that are provided by the model user; (3) management of privacy and security of data access when connecting external data; (4) automated data pre-processing. Data providers can offer an enhanced smart service: data no longer needs to be uploaded to the digital platform in its entirety; instead, the data provider can offer it into a federated workflow that is executed by the digital platform in return for value transfer. This makes it possible for the user to access larger data pools for training, to conduct transfer learning directly on the platform, and to access compute resources without having to establish their own infrastructure. The increased value exchange is to be captured via additional revenue streams: the market operator can charge additional fees for model training or for providing access to third party data that is not held on the platform; the platform provider can secure revenue from computational resources used for model training; and data providers can more closely monitor and monetise the use of their data in transactions.

Consumer load curve forecasting is a use case that is well suited to this archetype. The best performing models need to be retrained on data that representative of the target. The dynamic AI marketplace provides an enabling infrastructure that could support model personalisation, host anonymised datasets, and enable dataset joins. Web3 technology may also be used to manage data contracting or secure data during use in training (Fig. 3).

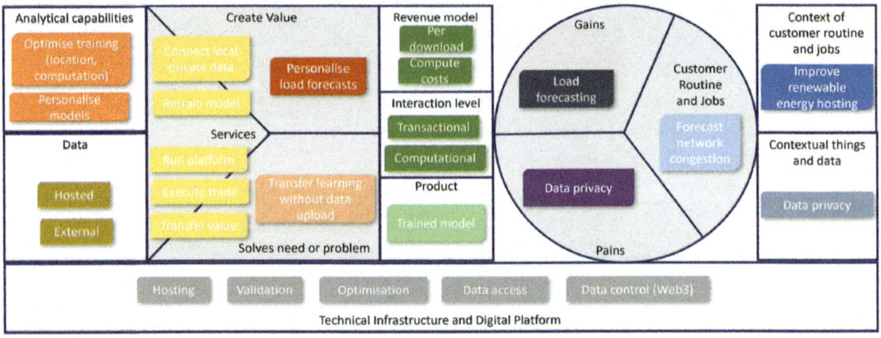

Fig. 3 Example dynamic AI marketplace for load curve prediction

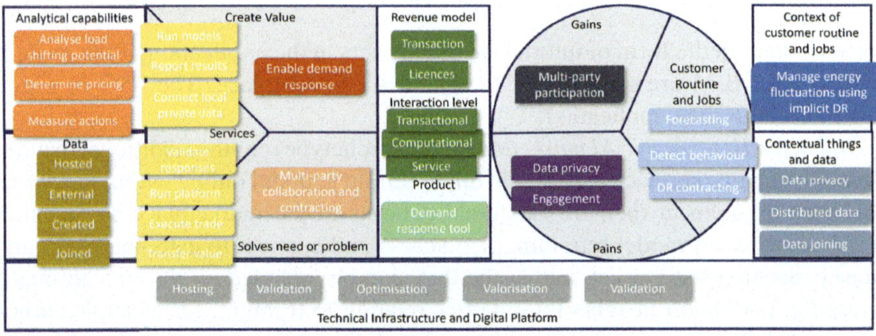

Fig. 4 Example AI marketplace for digital solutions

Archetype III: Solution AI marketplace. The third archetype further extends the dynamic marketplace by integrating models and data into additional abstraction layers (for example through hosted applications), allowing new user types to derive value from the platform, and increasing the revenue potential for model or data providers. Applications integrate and make use of data infrastructure, digital products and services already deployed in Archetype II. The marketplace also enables solutions based on newly generated or streamed data that is accessed via secure or federated protocols. The SSC in Fig. 4 illustrates the main features of the solution AI marketplace in relation to a DR workflow that involves: accessing data from household energy consumption while complying to data privacy regulations; executing models for prediction and reporting results to the programme manager; communicating consumption flexibility targets to users (e.g., set points and prices); applying NILM to determine demand side actions; facilitating multi-party contracting to allow value to be exchanged between all parties; and managing trade of value in response to user actions.

4 Conclusions and Next Steps

In this work, we develop business models for digital services in AI marketplaces, mapping the value proposition to the system value chain is the core of our methodology. The static marketplace is essentially a repository for digital assets and a platform where providers and users meet to access data and models. The dynamic marketplace allows in addition model retraining and the connection of external datasets. The solution marketplace envisions to create applications on top of the dynamic marketplace, enabling final users to securely participate in DR programs. Upon implementation and testing of these archetypes, the business models may evolve organically, but we can steer them with further iterations of the design steps we developed, for example in workshop sessions including the main stakeholders.

The commercial realisation of AI marketplaces is underway in many industry sectors as they can become better tools for value co-creation by: reducing redundant storage and computing, thus carbon footprint (avoiding retraining or duplication of model development); exploiting data under different ownership and regulatory frameworks; offering a less centralised alternative to data hubs; integrating many blocks of the value chain, for example by asking data providers to perform additional tasks like labelling. Clearly, there are also many business and technical challenges. Our next steps revolve around offering the final users of Archetype III control over their data that complies with different data privacy regulations; and liaising with DSOs, OEMs, and data subjects to automate access to data from smart meters. These will support federated learning approaches that in turn offer a reduction of data transfer volumes.

Acknowledgements This research is supported by the Swiss Innovation Agency via the NTN Innovation booster Energy Lab.

References

Berkhout, V., Frey, C., Hertweck, P., & Wickert, M. (2022). Chapter 20 Energy data space. In B. Otto, M. Ten Hompel, & S. Wrobel (Eds.), *Designing data spaces: the ecosystem approach to competitive advantage*.

Crofton, K., Wanless, E., & Wetzel, D. (2015). *The Electricity System Value Chain*. Rocky Mountain Institute: Technical report.

Gopstein, A., Nguyen, C., O'Fallon, C., Hastings, N., & Wollman, D. (2021). NIST framework and roadmap for smart grid interoperability standards, release 4.0. Technical Report NIST SP 1108r4, National Institute of Standards and Technology, Gaithersburg, MD.

Hussain, S., Gustavsson, R., Saleem, A., & Nordstrom, L. (2014). SLA conceptual framework for coordinating and monitoring information flow in Smart Grid. In *ISGT 2014* (pp. 1–5). IEEE.

Kern, D., Ensinger, A., Hammer, C., Neufeld, C., Lecon, C., Nagl, A., Bozem, K., Harrison, D. K., & Wood, B. M. (2022). Application Possibilities of Artificial Intelligence in a Renewable Energy Platform. In S. West, J. Meierhofer, & U. Mangla (Eds.), *Smart Services Summit* (pp. 35–43). Progress in IS: Springer International Publishing. Series Title.

Kumar, A., Finley, B., Braud, T., Tarkoma, S., & Hui, P. (2021). Sketching an AI marketplace: Tech, economic, and regulatory aspects. *IEEE Access,* 9, 13761–13774. This document https://arxiv.org/abs/2003.01593 seems to be an earlier version or preprint.

Mehri, V. A. (2019). *Towards secure collaborative AI service chains*. Blekinge Institute of Technology.

Mehri, V. A. & Tutschku, K. (2017). Flexible privacy and high trust in the next generation internet—the use case of a cloud-based marketplace for AI. In *Swedish national computer networking workshop* (p. 5). Halmstad University.

Ouyang, L., Yuan, Y., & Wang, F.-Y. (2022). Learning markets: An AI collaboration framework based on blockchain and smart contracts. *IEEE Internet of Things Journal, 9*(16), 14.

Poeppelbuss, J., & Durst, C. (2019). Smart service canvas—A tool for analyzing and designing smart product-service systems. *Procedia CIRP, 83,* 324–329.

Prado, M. D., Su, J., Saeed, R., Keller, L., Vallez, N., Anderson, A., Gregg, D., Benini, L., Llewellynn, T., Ouerhani, N., & others (2020). Bonseyes AI pipeline-bringing AI to you: End-to-end integration of data, algorithms, and deployment tools. *ACM Transactions on Internet of Things, 1*(4), 1–25.

Singh, M., Jiao, J., Klobasa, M., & Frietsch, R. (2022). Servitization of energy sector: Emerging service business models and startup's participation. *Energies, 15*(7), 2705.

Verdugo Cedeño, J. M., Hannola, L., & Ojanen, V. (2019). Knowledge requirements for sustainable smart service design:. In *Proceedings of the 11th International joint conference on knowledge discovery, knowledge engineering and knowledge management* (pp. 195–202). SCITEPRESS - Science and Technology Publications.

West, S., Gaiardelli, P., & Rapaccini, M. (2018). Exploring technology-driven service innovation in manufacturing firms through the lens of Service Dominant logic. *IFAC-PapersOnLine, 51*(11), 1317–1322.

Potential for the Development of Service Business Models Through Digital Services and the Active Inclusion of Customers

Gloria Bohn, Julius Taubert, Maximilian de Geus, Stefan Schweiger, and Valerie Bass

Abstract As one of the most important branches of the industry in Germany and the European Union, the mechanical and plant engineering sector is confronted with fundamental changes due to ever shorter innovation cycles and increased competitive pressure. This makes it even more important to increase the level of service components in business models with a low service level, which are still frequently found in SMEs. This paper is dedicated to the changes that the individual components of a business model have experienced and will experience. Special attention is paid to economic sustainability, since service business models can also positively influence the long-term nature of a business. Seven interviews conducted with relevant companies serve as the empirical basis of this paper. The analysed effects of smart services and active customer integration are structured and summarized within the three pillars of every business model (value proposition, the value creation architecture and the revenue mechanic).

Keywords Customer integration · Knowledge management · Service business model · Smart services · Digitalization · Small and medium-sized enterprises

1 Introduction

The plant and mechanical engineering sector is one of Germany's largest industries, accounting for 3.4% of GDP in 2019 (Federal Statistical office & of Germany: Gross value added in Germany in the years, 2009–2019). For decades, technological leadership and innovative ability were seen as guarantees of success in this industry (FTI-Andersch: Welche Herausforderungen deutsche Maschinenbauer jetzt meistern müssen., 2022). Meanwhile, shortened innovation cycles as well as increasing international competition and the associated pressure on prices are putting capital

G. Bohn · J. Taubert · M. de Geus · S. Schweiger (✉) · V. Bass
HTWG Konstanz—University of Applied Sciences, Alfred-Wachtel-Str. 8, 78462 Konstanz, Germany
e-mail: smart-service-bw@htwg-konstanz.de

J. Meierhofer et al. (eds.), *Smart Services Summit*, Progress in IS,
https://doi.org/10.1007/978-3-031-36698-7_9

goods manufacturers under severe pressure. The spotlight of innovation is therefore no longer on new product developments only. Product-related services are becoming increasingly important as a differentiating factor in international competition. The Fourth Industrial Revolution and the digital transformation will have a huge impact on the future of the companies in this industry. Digital services increasingly offer opportunities to collaborate more closely with customers.

This makes it possible to rethink the architecture of business models in order to ensure sustainable business success from an economic perspective. Digital services and the active involvement of customers offer new potential for the value proposition, the value creation architecture and the revenue mechanics. Especially with regard to the value proposition and the value creation architecture, interactive value creation can generate added value both on the customer side and on the supplier side and hence increase the economic sustainability of companies in this industry.

In this context, this paper answers the question how service business models change by the involvement of customers and which opportunities can arise in particular for SMEs of the plant and mechanical engineering sector through the active involvement of customers through digital services. Furthermore, it aims to examine the connection between service business models and long-term economic sustainability.

2 Literature Review

First, some important definitions are presented at this point. Small and medium-sized enterprises (SMEs) served as the object of investigation for the empirical study. According to EU Recommendation 2003/361, representatives of this group are enterprises with no more than 250 employees or an annual turnover of up to 50 million euros or a balance sheet total of up to 43 million euros (Commission recommendation of 6 May, 2003 concerning the definition of micro, small & medium-sized enterprises 2003).

This text also deals with business models. Numerous definitions exist here, as there is no consensus in academia on a universally valid definition. For this paper, the definition approach according to Osterwalder and Pigneur was used: "A business model describes the rationale of how an organisation creates, delivers, and captures value" (Osterwalder, 2010).

Based on the underlying structural design of business models, the concept of interactive value creation can be anchored. According to this, customers do not only act as passive recipients of services or goods, which the mere definition of a business model would suggest. Rather, in the context of interactive value creation, customers represent value creation partners of companies e.g. by helping to shape products and services, whereby in the context of this value creation, customers and providers are not completely individual and independent actors (Rusthollkarhu et al., 2021). This transforming role of the customer also changes the logic of interaction as well as the business model design, whereby the type of value creation is no longer dominated by the provider (Reichwald & Piller, 2009).

Directly related to interactive value creation is the so-called Service Dominant Logic (SDL), which was specified by Vargo and Lusch for industrial companies (Husen, 2015). Through the integration of services and tangible goods, goods undergo a further development from end products to objects for services in a value creation process, whereby the focus shifts from the original sale of tangible goods to an exchange of intangible skills, information and knowledge (Vargo & Lusch, 2016) (Kalkhofer et al., 2021). The idea of interactive value creation is thus concretized and further developed by the concept of S-D logic, whereby this focuses on the process of serving and not on the pure output (Lush & Nambisan, 2015) (Burton et al., 2021).

At the present time most companies are aware that a purely technological lead can hardly bring competitive advantage and are continuously trying to expand the service business (Altenfelder et al., 2021). However, various barriers exist: On the one hand, the associated digital solutions have so far only represented isolated solutions which is the result of a lack of industry standards. On the other hand, many companies in the mechanical and plant engineering sector have a rather low level of digital maturity which impairs the establishment of sustainable and innovative service business models. Instead, the strategic orientation is still strongly product-driven and focused on the new machinery business. This makes it possible to rethink the architecture of business models. Digital services and the active involvement of customers offer new potential for the value proposition, the value creation architecture and the revenue mechanics. Particularly with regard to the value proposition and the value creation architecture, interactive value creation can create added value both on the customer side and on the company side which is outlined in this paper.

3 Methodology

In order to gain knowledge about the current situation of SMEs of the plant and mechanical engineering sector regarding service business models, expert interviews were conducted with representatives of SMEs. In total, seven representatives from service management, IT, business development and sales have been interviewed. Further criteria for the choice of interviewees were the company size, location in Germany, plant and mechanical engineering focus and experience with service business models. The interviews were open-ended, semi-structured and from a retrospective perspective which also included the discussion of target images. Subsequently, the interviews were evaluated in a qualitative content analysis.

Within the framework of the empirical study presented in this text, companies with a maximum number of 2000 employees were also included, thus deviating from the EU recommendation (Born, 2018). The authors justify this by stating that it is predominantly important for the empirical study that the companies already have experience with service business models while at the same time having a medium-sized company structure and processes.

In a next step, the interviews were line-by-line coded and then first and second order categories were formed. This resulted in a codebook with a total of 131 codes that reflect the answers from the guided interview. The codes were further condensed and the frequency of the coded text passages evaluated according to Gioia methodology (Gioia et al., 2013).

4 Results

If one wants to investigate potentials for service business model development through digital services and the active involvement of customers, it makes sense to take a closer look at the value proposition, the value creation architecture and the revenue mechanics. The following Fig. 1 shows a basic structure on which sustainable corporate success is built.

4.1 *Potentials for the Value Proposition*

Service activities of the companies surveyed were mostly limited to spare parts, maintenance or repairs. From the survey, it can be deduced that there is an increasing trend from the classic machinery and plant manufacturer to the solution provider. This includes more comprehensive customer support through upstream and downstream service processes. Companies recognize that advancing digitalization offers numerous opportunities in the value proposition and in the collaboration with customers. When developing new service offerings, the companies surveyed pay close attention to ensuring that these are customer-centric services, creating direct added value for the customer. The most frequently mentioned digital services that are currently being developed by the companies and offered to the customer are e.g.

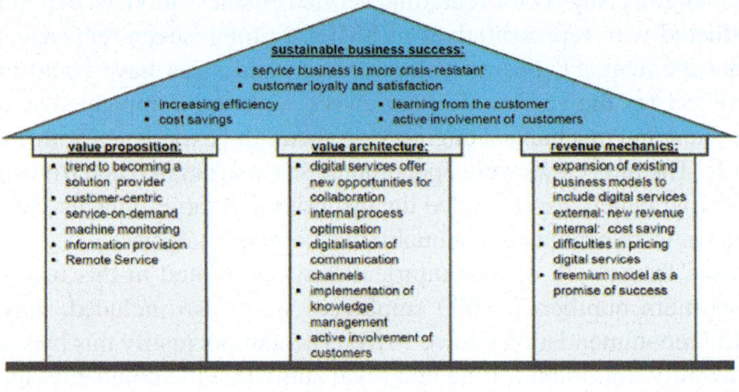

Fig. 1 Structure of sustainable business success

service-on-demand, machine monitoring and the provision of information or rather knowledge through a customer portal. Remote services which can intensify fast and direct customer contact are a fixed component of smart services that was mentioned repeatedly by the companies surveyed. Machine monitoring enables the customer to view various machine data in real time. In this context, it is also possible to evaluate the customer's machine data on request in order to make numerous forecasts or controls in the subsequent step. In addition to the daily business in the context of production activities, digital solutions also noticeably change the training offered of the companies surveyed. Thus, the majority of the interviewed companies are currently switching from classroom training to online training, if they have not already done so. Numerous added values can be derived from the listed digital services for both the provider and the customer. The greatest and most frequently mentioned added value lies in higher machine availability, due to faster response times and a faster flow of information. Consequently, the improved value proposition contributes to increased efficiency. Furthermore, it leads to higher competitiveness due to cost savings. The provision of a customer portal or joint information platforms makes it easier for the customer to always be able to access relevant information as needed. Through self-service or quick contact with the right expert, less experienced employees on the customer side are needed to carry out problem analyses and repairs or maintenance independently. This represents great added value, especially in regard of the shortage of skilled workers across all sectors. In addition, the companies recognize the opportunity to promote environmental sustainability through a noticeable reduction in business travel and the extension of product life cycles.

4.2 Potentials for the Value Architecture

The potentials for the value proposition through smart services leads in a subsequent step to a change in the value architecture. Thus, digital services and information platforms offer companies new opportunities to collaborate with their customers. However, this new form of cooperation can only succeed through an active rethinking, away from the pure machinery provider towards interactive value creation. Customers should not only be passive recipients of services but take an active role in the value creation. The resulting change in the value creation architecture also changes the traditional business model structures.

For this transformation process to succeed, according to the companies surveyed, it is first necessary to revise their own processes and resources before customers can be actively and meaningfully integrated into the value creation. The main tasks in changing the value creation architecture of the companies surveyed are to digitalize their own service processes. Here, the focus is particularly on the digitalization of communication channels and document management systems. The companies interviewed frequently mentioned the introduction of a ticket system, a customer portal and the offering of video calls in order to be able to contact customers quickly and competently. The design of paperless work processes and the establishment of a knowledge management system were mentioned here with similar frequency. Knowledge-centered work design makes it possible to generate service-relevant knowledge about machines and plants together with customers. A mutual added value arises in preventing previous information asymmetries that the customer has carried out repairs or maintenance without the knowledge of the plant engineers. Conventional information asymmetries have massively hindered efficient service and resulted in the manufacturer being hid of crucial information or rather knowledge on weak points and potential improvements to the product. Thus, some of the companies surveyed are already using the possibility of joint maintenance of machine files and close cooperation with the customer to adapt development and design activities as a result. Therefore, the result of the changed value creation architecture is that an active involvement of the customer is a great advantage in the context of product development with a targeted focus on customer benefits which ultimately has a positive effect on the two-sided value creation architecture.

4.3 Potentials for Revenue Mechanics

Building on the changed value creation architecture, the potentials in the revenue mechanics should be illuminated at this point. The expansion of the business model to include digital services and the active integration of customers offer new potential in the revenue mechanics. The revenue mechanics include new revenue streams on the external side and changes in the cost structure on the internal side. From the interviews conducted, it can be seen that the focus is currently on internal cost reduction and increasing efficiency, as a large part of the digital value proposition and the digital collaboration platforms are used within the framework of warranty phases or are also offered free of charge. As reflected in the interviews, pricing digital services is often difficult due to insufficient experience in designing digital business models and reference values in the market. Analysis results of the companies surveyed, which were reported during data collection, show that freemium models, in which services are initially offered free of charge for a certain period of time and can then be used for a fee, are seen as promising. Furthermore, stable revenue expansion can be achieved by integrating customers into a contract system (e.g. maintenance contracts). In contrast to the revenue side, the interview participants state that the greatest profit is achieved through internal efficiency increases. Especially during the warranty phase, service interventions are directly detrimental to the profit margin of the new machine business. Remote service and the active involvement of customers in repair or maintenance work represent measures to optimize service calls and increase the profit margin by reducing costs. The mentioned reasons for this are to promote economic and ecological sustainability by reducing travel and freeing up the capacities of service technicians for more profitable activities. Parallel to a more efficient use of the scarce resources of service technicians, a closer cooperation with the customer accelerates and improves the process of problem identification and troubleshooting, whereby subsequent activities can be carried out in a more targeted and faster manner.

4.4 *Impact on Sustainable Business Success*

As the survey shows, in addition to generating new revenues and cost savings, efficiency gains create further potential that can contribute to sustainable business success. Particularly valuable for the companies surveyed is the more intensive relationship with customers which can only grow effectively with the use of digital communication channels and shared knowledge platforms. A higher level of collaboration with customers, among other things through the networking of systems and the associated direct exchange of data, confirms the companies' statement. The thesis behind this is "learning from customers" and opens the possibility for SMEs of the mechanical engineering sector in particular to benefit from the development and research activities of their mostly larger customers. The customers make it possible for the SME-based plant manufacturers to gain access to know-how and human resources that enable the joint advancement of optimization potentials. The companies surveyed observe that this development has a beneficial effect on the frequently mentioned shortage of skilled workers in the industry, as knowledge management enables customers to carry out individual service tasks independently or under remote supervision, thus relieving their own employees and increasing efficiency in the deployment planning of scarce resources. Because solution providers can bind their customers more strongly to a company than pure product providers, companies in the target industry can ensure more sustainable business success. Especially in the sector of small and medium-sized machine and plant manufacturers, customer loyalty plays an essential role, since service success and customer satisfaction are the fundamental basis for repurchase rates by customers. The companies surveyed hope to gain significant competitive advantages from the aspects listed so far. The majority of the companies surveyed have confirmed that investments in sustainable service business models make economic sense, as the service business is crisis-resistant and can compensate for fluctuations in the new machine business.

5 Discussion

The research allows implications for the economy as well as for science.

Economic implications: This study carved out that SMEs of the plant and mechanical engineering sector perceive that digital services hold several opportunities for the development of business models and the inclusion of and collaboration with customers across the value chain which leads to benefits and competitive advantage on both customer and producer side. Such benefits are e.g. the reduce of information asymmetry which used to hamper efficient services and a common knowledge base as well as the improvement of the company's economic and ecological sustainability. The latter effect can be explained through the stable sales expansion by including customers in a long-term contract system, raising profits through increased efficiency and reduced costs and new opportunities to increase customer retention all enabled by the introduction of digital services.

Scientific implications: From the scientific view, this paper contributes insides into the current situation of the plant and mechanical engineering sector in Germany regarding the development of digital services and the effect on a two-dimensional collaboration between companies and their customers. As pointed out, the companies start to change their product-centered view into a service-centered view. However, the interviewed companies are at the beginning of this development and face several obstacles. Therefore, there are many fields for future research to deepen the results of this study. For example, the pricing of digital services was mentioned as difficulty and needs to be addressed further. Furthermore, this paper pointed out how digital services and the active inclusion of customers can foster economic and ecological sustainability.

6 Conclusion

The research question how service business models change by the involvement of customers and which opportunities can arise in particular for SMEs of the plant and mechanical engineering sector through digital services were answered in this paper. It can be stated that the interviewed companies are at the beginning of this development and digital services still hold many opportunities to improve customer integration and in turn competitive advantage. Furthermore, this paper concluded numerous ways on how this development can increase a company's economic and ecological sustainability. In conclusion, the development of digital services and the inclusion of customers in the value creation is an important task for companies of the plant and mechanical engineering sector that science and economy should focus on in the future.

References

Altenfelder, K., Schönfeld, D. and Krenkler, W., eds. (2021). *Services management und digitale Transformation: Impulse und Beispiele für die erfolgreiche Umsetzung digitaler Services.* Springer Gabler

Born, H.-J. (2018). *Geschäftsmodell-Innovation im Zeitalter der vierten industriellen Revolution: Strategisches Management im Maschinenbau.* Springer Vieweg. [Online]. http://www.springer.com/

Burton, J., Raddats, C., Story, V., West, S. and Zolkiewski, J. (2022). Moving beyond manufacturing: building a research agenda for servitizing service firms and ecosystems in *Smart Services Summit: Smart Services Supporting the New Normal, 2021*, 149–158

Commission Recommendation of 6 May 2003 concerning the definition of micro, small and medium-sized enterprises OJEU. (2003). *Official Journal of the European Union*

Federal Statistical Office of Germany: Gross value added in Germany in the years 2009 to 2019. https://de.statista.com/statistik/studie/id/6374/dokument/maschinenbau-in-deutschland/

FTI-Andersch: *Welche Herausforderungen deutsche Maschinenbauer jetzt meistern müssen.* (2022). https://www.fti-andersch.com/de/insights/welche-herausforderungen-deutsche-maschinenbauer-jetzt-meisternmuessen/

Gioia, D. A., Corley, K. G., & Hamilton, A. L. (2013). Seeking qualitative rigor in inductive research. *Organizational Research Methods, 16*(1), 15–31. https://doi.org/10.1177/109442811245215

van Husen, C. (2015). Neue Serviceprodukte in industriellen Wertschöpfungsnetzwerken. In M. Bruhn & K. Hadwich, (Eds.) *Interaktive Wertschöpfung durch Dienstleistungen* (pp. 493–515). Springer Fachmedien Wiesbaden

Kalkhofer, H., Moosbrugger, N., Ulmer, A. & Dobler, M. (2022). Enable service interaction and value co-creation for small and medium-sized enterprises through an innovation-method-framework in *Smart Services Summit: Smart Services Supporting the New Normal*, 2021, 127–136.

Lush, R. F. & Nambisan, S. (2015) Service Innovation: a service-dominant logic perspective. *MIS Quarterly*, 155–176

Osterwalder, A. (2010). *Business model generation: A handbook for visionaries game changers and challengers.* Wiley (2010). [Online]. https://learning.oreilly.com/library/view/-/9780470876411/?ar

Reichwald, R., & Piller, F. T. (2009). *Interaktive Wertschöpfung: Open Innovation, Individualisierung und neue Formen der Arbeitsteilung* (2nd ed.). Gabler Verlag/GWV Fachverlage GmbH Wiesbaden.

Rustholkarhu, S., Hautamaki, P., & Aarikka-Stenroos, L. (2021). Value (co-)creation in B2B sales ecosystems. *JBIM, 36*(4), 590–598. https://doi.org/10.1108/JBIM-03-2020-0130

Vargo, S. L., & Lusch, R. F. (2016). Institutions and axioms: An extension and update of service-dominant logic. *Journal of the Academy of Marketing Science, 44*(1), 5–23. https://doi.org/10.1007/s11747-015-0456-3

Smart Service Development in Entrepreneurship and Intrapreneurship Support Infrastructures in Vorarlberg and the Lake Constance Region—Status Quo and Framework for Practitioners

Martin Dobler and Magdalena Meusburger

Abstract The role of entrepreneurs and intrapreneurs in the current zeitgeist is to drive innovation, re-shape rigid, established processes in business as well as for consumers. They use new viewpoints to pioneer new (business) models which focus on 'smartness' rather than the purely monetary and short-sighted models of yesteryear. Fostering and supporting the culture of this current zeitgeist is a mayor challenge for entre- and intrapreneurial support infrastructures, namely startup centres and innovation hubs of universities and other public institutions as well as innovation centres of private companies. Hereby, support may range from access to funding over provision of resources such as offices or computing hardware to coaching in the development of business ideas and strategic roadmaps for product and service deployment. In this paper, we focus on describing the status-quo of aforementioned support infrastructures in Vorarlberg and the Lake Constance region, then extend the scope to existing (international) approaches for aiding founders and innovators in the development of smart services. An analysis of success stories of the Vorarlberg startup centre 'startupstube' and other initiatives including their comparison to international counterparts builds the basis for a methodological framework for (service science) coaching in entre- and intrapreneurial support infrastructures. The paper is concluded by the description of a framework for choosing the right methods and tools to create service value in entre-/intrapreneurship based upon tested, proven know-how and for defining support infrastructure needs based upon pre-defined stakeholder and target groups as well as the (industry) sectors of the innovators.

M. Dobler (✉)
Vorarlberg University of Applied Sciences, Research Centre Business Informatics, Hochschulstraße 1, Dornbirn, Austria
e-mail: martin.dobler@fhv.at

M. Meusburger
Department of Business Management, Vorarlberg University of Applied Sciences, Hochschulstraße 1, Dornbirn, Austria
e-mail: magdalena.meusburger@fhv.at

J. Meierhofer et al. (eds.), *Smart Services Summit*, Progress in IS,
https://doi.org/10.1007/978-3-031-36698-7_10

Keywords Entrepreneurship · Intrapreneurship · Smart services · Smart service development

1 Introduction

Since 2010, entrepreneurs in Austria founded over 2,800 startups (77 in Vorarlberg) of which 22% are the result of academic settings and their most common background is in technical studies or research (AIT Austrian Institute of Technology, 2021). Entre- and intrapreneurship support infrastructures play an important role in aiding founders with access to funding, development of sustainable product and business ideas, as well as ensuring their market fit (Kasouf et al., 2008). However, currently a sound approach for the support of servitisiation and implementation of smart services is missing. Therefore, we analyse the status quo and propose a smart service framework for practitioners in entre-/intrapreneurial support infrastructures.

Whereas services play an important role in the process of business development, Smart Services and Service Dominant Logic (SDL) might be considered an afterthought by founders, the support portfolio in founding centres (entrepreneurship) and innovation centres (intrapreneurship) (Lindhult et al., 2018). The same applies even in the zeitgeist conforming role definition of innovators, i.e., the concept of the 'Innopreneur' (Gündoğdu, 2012). Service science in theory (Kindström & Kowalkowski, 2014; Kowalkowski & Ulaga, 2017) and its application in practice (ZHAW, 2020) is defining a number of processes for supporting companies, e.g. Product Lifecycle Services (PLS), Asset Efficiency Services (AES) and Process Supporting Services (PSS). Whilst these processes and the eight foundational premises of SDL's service-centred view (Vargo & Lusch, 2014) are implied in support infrastructures for entrepreneurship and intrapreneurship, ranging from coaching over methodological toolboxes and know-how to co-creation approaches in quadruple helix constellations, a clearer model for aiding entrepreneurs and intrapreneurs holistically is needed to deliver a service science perspective on business model innovation (Maglio & Spohrer, 2013).

2 Research Design

To determine the composition of the current status quo of service science in entre- and intrapreneurial support infrastructures, we conduct a desk and literature research on the approaches towards teaching service science, applying it in various settings in teaching, coaching, and for consulting, as well as classification of service sciences concepts for various application settings. Furthermore, we analysed conducted supporting activities of the Vorarlberg University of Applied Sciences' startup centre 'startupstube' and present a framework for selection of methods for supporting value creation in entre-/intrapreneurship settings. For intrapreneurship, we analysed results from the programme KMUdigital and the project Data Sharing Framework; in entrepreneurship, we used coachings, workshops, talks, and founder stories of the startup centre 'startupstube', which has coached over 150 founding teams with startup ideas and held 50+ events with more than 3200 participants in the past five years.

For ease of categorisation and usability, we use seven common scenarios (see Chap. 4) as starting points for innovation support activities, all derived from above approach.

3 Smart Services Development for Innovators: State-Of-The-Art and Literature Research

3.1 Methods, Smart Service Development, and Stakeholders

Methods for supporting, selecting, and assessing sustainable (social, economic, and environmental) innovation are countless. These range from methods to assess the innovation potential of (micro-)SMEs (e.g. via ranking of suppliers (Gupta & Barua, 2018) or evaluation of eco-innovation assessment (Marin et al., 2015)) over data/indicator driven approaches (e.g. tools as Crunchbase Żbikowski & Antosiuk, 2021 and others) to traditional methods of business model development. The latter deal with questions on how to explain the business, how to run the business, and how to develop the business (Spieth et al., 2014). In recent years, the methods have also been extended to questions on how to make business models and their inherent innovation more sustainable (Geissdoerfer et al., 2018). Sustainability, hereby, might address the (monetary) sustainability of ideas and business impact (Saebi et al., 2017) and/or eco-innovation, especially when taking the current trend of Circular Economy into account (Vence, 2019). As it is probably true for the success of all pioneering ideas, Carrillo-Hermosilla et al. (2009) argue that (eco-)innovation is mainly driven by technological development and, importantly, by organisational change and innovation mindset of groups and individuals.

When analysing the set of available methods, it becomes clear that innovation methods and its related approaches all operate in multi-criteria problem spaces; a fact which has also been addressed in meta-analysis of innovation characteristics from the earliest stages of research in the field (Tornatzky and Klein, 1982).

To tackle the aforementioned inherent complexity and manage the needs and potentials of the ongoing trend towards servitisation, the integration of SDL approaches should be taken into consideration. In this paper's targeted region and Europe in particular, a movement towards servitisation is noticeable, albeit not exploited to its full potential yet (Dachs et al., 2014).

Notable methods are based upon the eight foundational principles and four process types commonplace in SDL. Extended processes for supporting companies include Product Lifecycle Services (PLS), Asset Efficiency Services (AES) and Process Supporting Services (PSS) (Kindström & Kowalkowski, 2014; Kowalkowski & Ulaga, 2017; ZHAW, 2020). Specific extensions for traditional business plan modelling using SDL have been made for the Osterwalder Business Model Canvas (Ojasalo & Ojasalo, 2018), incorporated into Design Thinking (Geissdoerfer et al., 2016), and further holistic approaches (Clauß et al., 2014).

As SDL, business model development and entrepreneurial support is driven by creating value for stakeholders. Besides value creation for customers and shareholders, innovation has long been known to support a wide variety of stakeholders, local and regional development and societal prosperity through industrial, educational and governmental structures (Kindström & Kowalkowski, 2014). However, the innovative and sustainable development of a country depends not only on the presence of a competent public sector, universities and industries, but more importantly on how they work together to achieve their own strategic goals. Using regional and local capacities, human capital is thus placed at the centre of innovation.

A key concept that can be applied to foster the innovation potential of an ecosystem is the Quadruple Helix model. The Quadruple Helix innovation model is an innovation collaboration model that links academia, business, government, and society together in the development of innovations. The model introduces additional participants in innovation cooperation to allow for a wide range of possible actors such as civil society organisations, citizens, collaborators, and others. In innovation settings which are driven predominantly by the public sector, the Quadruple Helix might be replaced by a Public–Private(-People)-Partnership (PPP, PPPP) model. (Meissner, 2019).

To round of our selected choices of methods, we add Co-Creation (Sanders & Stappers, 2008) as a separate class of methods, i.e., distinct from the application of Quadruple Helix ideas. The distinction becomes necessary since Quadruple Helix approaches require a pre-determined set of actors or stakeholder groups and Co-Creation methods can be applied if one group is missing or even in teams of individuals, e.g., team-based innovation workshop in intrapreneurial settings within a company.

For purely entrepreneurial innovation, we consider the Lean Startup method (Eisenmann et al., 2012).

4 Findings and Examples of Applied Support for Service Design

In essence, the entrepreneurial process consists of the identification and exploration of entrepreneurial opportunities (Shane, 2003). Based upon a detailed analysis of supporting activities of the startup centre 'startupstube', the programme KMUdigital and the project Data Sharing Framework, as well as the local innovation ecosystem in general, the following scenarios were identified (entrepreneurial examples in brackets).

1. Idea-based approach: An aspiring entrepreneur identifies an opportunity in the marketplace and develops a startup idea. The entrepreneur gathers a team and or other stakeholders supporting the idea and together they enter the user validation phase and are looking for problem–solution fit (*Example: Die Limomacher, tree.ly*).
2. Resource-based approach: A team of aspiring entrepreneurs or a solo entrepreneur use an effectual logic (Becker et al., 2015) and look at who they are, what they know, and who they know to determine the means at their disposal to develop their startup idea (*Example: Kräuternest*).
3. Research-based approach: Novel research results are the initial point of the startup journey. A team of aspiring entrepreneurs or a solo entrepreneur identify possible applications and opportunities in the marketplace (*Example: SANlight*).
4. Intrapreneurship resource-based approach: A company or public institution establishes a dedicated team with a mandate to identify and develop startup or spin-off ideas (*Examples: Plattform V, Greenchips*).
5. Intrapreneurship idea-based approach: A company or public institution sources startup ideas from employees and other stakeholders, selects the ideas based on organizational criteria and assembles a team to make the idea workable (*Example: Weavs GmbH*).
6. Social entrepreneurship approach: A social challenge or problem forms the starting point of the entrepreneurial journey. The team, or the solo entrepreneur is driven towards entrepreneurship by a dismal picture of existing divide (*Examples: AlpSIB, Fempower Community*).
7. Sustainable entrepreneurship approach: The vision of the entrepreneur or the team to realise a sustainable innovation aimed at the mass market (Schaltegger & Wagner, 2011) and taking into account ecological, economic and social issues, kicks off the startup process (*Example: Luculla Culinaria*).

The different scenarios—upon which the following framework has been developed—each require a specific variety of methods and tools to support them along their entrepreneurial process. There is no *One-Size-Fits-All* approach to supporting them (Bass & Taubert, 2022).

5 Train-The-Trainers: Framework for Service Value Creation in Entre- & Intrapreneurship

Based the scenarios, we derive the framework for practitioners as shown in Fig. 1. The framework is intended to be used by coaches in support infrastructures and might also be used to teach coaches different ways to tackle common and reoccurring support requests (*train-the-trainers*). The framework is split up into the phases of

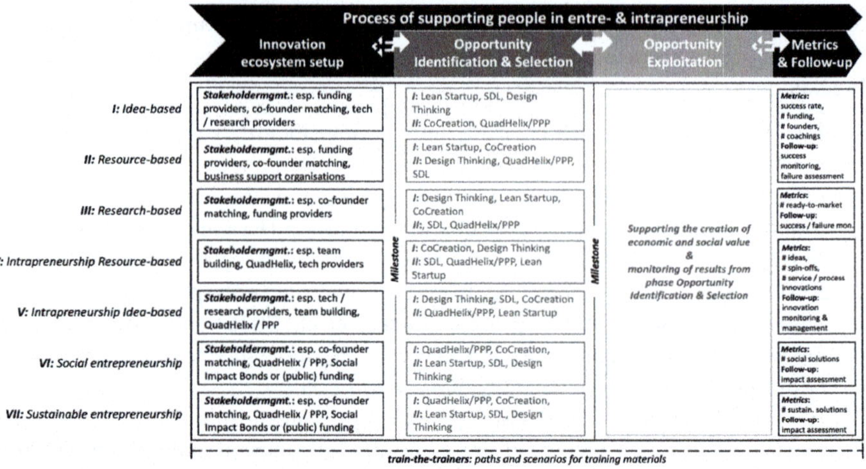

Fig. 1 Service innovation framework for practitioners in support infrastructures

the support processes: innovation ecosystem setup, opportunity phases, as well as a final phase for collecting indicators and conducting follow-up measures. For each of the phases, we list a prioritisation of the respective measures, methods, metrics or follow-up actions. Method prioritisation is split into primary and secondary methods (indicated as I and II in Fig. 1). It is important to notice that the prioritisation is intended to be used as a guideline, as the final decision on the instruments and tools to use is always taken by the coach and their understanding of the innovation idea and the created service value.

6 Summary and Outlook

In this paper, we have presented existing methods and approaches to support intra- and entrepreneurial activities, including methods from smart service development, Design Thinking, and Co-Creation approaches—extended by approaches to tailor to and include stakeholder groups, namely Quadruple Helix approaches and PP(P)P. Based upon a real-world sample set of startups and support activities, we derive different scenarios and link them in a framework with the methodological approaches, pre-requisites, and metrics as well as follow-up measures.

In future, the continuation of our work will focus on improving *train-the-trainers* materials (i.e., describing best practices and real-word scenarios on how to support entrepreneurs) as well as assessing, describing and adapting follow-up measures (i.e., ensuring long-term success) and the long-term analysis of the metrics gathered (mainly measuring impact and sustainability indicators).

Acknowledgements This paper was partly made possible by the financial support of the EIT Manufacturing project 'EcoAction' (Action to Boost Ecosystem Impact through Cross-partner Learning; project n°: 943) and the Interreg V Alpenrhein-Bodensee-Hochrhein programme project 'Data Sharing Framework für KMU' (project n°: ABH097).

References

AIT Austrian Institute of Technology, Austrian Startup Monitor 2021, Home Town Media GmbH, ISBN 978–3–200008139–0

Bass, V., Taubert, J. (2022). Capability of Service Innovation—What SMEs Can Learn from Start-Ups. In: West, S., Meierhofer, J., Mangla, U. (eds) *Smart Services Summit. Progress in IS*. Springer, Cham. https://doi.org/10.1007/978-3-030-97042-0_3

Becker, A., Knyphausen-Aufseß, D. Z., & Brem, A. (2015). Beyond traditional developmental models: A fresh perspective on entrepreneurial new venture creation. *International Journal of Entrepreneurial Venturing, 7*(2), 152–172.

Carrillo-Hermosilla, J., González, P. R. D., & Könnölä, T. (2009). What is eco-innovation?. In *Eco-innovation* (pp. 6–27). Palgrave Macmillan

Clauß, T., Laudien, S. M., & Daxböck, B. (2014). Service-dominant logic and the business model concept: Toward a conceptual integration. *International Journal of Entrepreneurship and Innovation Management, 18*(4), 266–288.

Dachs, B., Biege, S., Borowiecki, M., Lay, G., Jäger, A., & Schartinger, D. (2014). Servitisation of European manufacturing: Evidence from a large scale database. *The Service Industries Journal, 34*(1), 5–23.

Eisenmann, T. R., Ries, E., & Dillard, S. (2012). Hypothesis-driven entrepreneurship: The lean startup. *Harvard Business School Entrepreneurial Management Case*, (812–095).

Geissdoerfer, M., Bocken, N. M., & Hultink, E. J. (2016). Design thinking to enhance the sustainable business modelling process–A workshop based on a value mapping process. *Journal of Cleaner Production, 135*, 1218–1232.

Geissdoerfer, M., Vladimirova, D., & Evans, S. (2018). Sustainable business model innovation: A review. *Journal of Cleaner Production, 198*, 401–416.

Gündoğdu, M. Ç. (2012). Re-thinking entrepreneurship, intrapreneurship, and innovation: A multi-concept perspective. *Procedia-Social and Behavioral Sciences, 41*, 296–303.

Gupta, H., & Barua, M. K. (2018). A novel hybrid multi-criteria method for supplier selection among SMEs on the basis of innovation ability. *International Journal of Logistics Research and Applications, 21*(3), 201–223.

Kasouf, C. J., Darroch, J., Hultman, C. M., & Miles, M. P. (2008). Service dominant logic: Implications at the marketing/entrepreneurship interface. *Journal of Research in Marketing and Entrepreneurship, 10*(1), 57–69.

Kindström, D., Kowalkowski, C. (2014). Service innovation in product-centric firms: A multi-dimensional business model perspective. *Journal of Business & Industrial Marketing, 29*(2), 96–111

Kowalkowski, C., Ulaga W. (2017). *Service strategy in action: A practical guide for growing your B2B service and solution business*. Service Strategy Press

Lindhult, E., Chirumalla, K., Oghazi, P., & Parida, V. (2018). Value logics for service innovation: Practice-driven implications for service-dominant logic. *Service Business, 12*(3), 457–481.

Maglio, P. P., & Spohrer, J. (2013). A service science perspective on business model innovation. *Industrial Marketing Management, 42*(5), 665–670.

Marin, G., Marzucchi, A., & Zoboli, R. (2015). SMEs and barriers to Eco-innovation in the EU: Exploring different firm profiles. *Journal of Evolutionary Economics, 25*(3), 671–705.

Meissner, D. (2019). Public-private partnership models for science, technology, and innovation cooperation. *Journal of the Knowledge Economy, 10*(4), 1341–1361.

Ojasalo, J., & Ojasalo, K. (2018). Service logic business model canvas. *Journal of Research in Marketing and Entrepreneurship, 20*(1), 70–98.

Saebi, T., Lien, L., & Foss, N. J. (2017). What drives business model adaptation? The impact of opportunities, threats and strategic orientation. *Long Range Planning, 50*(5), 567–581.

Sanders, E. B. N., & Stappers, P. J. (2008). Co-creation and the new landscapes of design. *Co-Design, 4*(1), 5–18.

Schaltegger, S., & Wagner, M. (2011). Sustainable entrepreneurship and sustainability innovation: Categories and interactions. *Business Strategy and the Environment, 20*(4), 222–237.

Shane, S. (2003). *A General Theory Of Entrepreneurship: The Individual-opportunity Nexus*. Edward Elgar.

Spieth, P., Schneckenberg, D., & Ricart, J. E. (2014). Business model innovation–state of the art and future challenges for the field. *R&d Management, 44*(3), 237–247.

Tornatzky, L. G., & Klein, K. J. (1982). Innovation characteristics and innovation adoption-implementation: A meta-analysis of findings. *IEEE Transactions on Engineering Management, 1*, 28–45.

Vargo, S. L., & Lusch, R. F. (2014). Evolving to a new dominant logic for marketing. In *The service-dominant logic of marketing* (pp. 21–46). Routledge.

Vence, X., & Pereira, Á. (2019). Eco-innovation and Circular Business Models as drivers for a circular economy. *Contaduría y administración, 64*(SPE1), 0–0.

Żbikowski, K., & Antosiuk, P. (2021). A machine learning, bias-free approach for predicting business success using Crunchbase data. *Information Processing & Management, 58*(4).

ZHAW Zürcher Hochschule für Angewandte Wissenschaften (Eds.). (2020). *Data4KMU Data Science für KMU leicht gemacht*. Aktuelle Erkenntnisse und Lösungen. IBH-Lab KMUdigital.

Emerging Lessons from Industry on How to Make Radical Innovation Sustainable

Shaun West and Dominic Boeni

Abstract This paper explores radical innovation in industrial firms in the DACH region to help identify success factors via an exploratory study. The underpinning assumption is that for real sustainability, firms must innovate for the long-term and deliver ongoing incremental innovation. The challenge is that different approaches are needed to be successful in both forms. In our study, we interview over 100 senior managers through a Delphi process to gain insights. We identify seven barriers and the cause of the obstacles, and we also recognise seven lessons that improve the sustainability of radical innovation. The synthesis of the work is that firms must become learning organisations and that they need to consider the behavioural aspects of management, the necessary adaptability of the process, and the long-term funding aspect of radical innovation.

Keywords Radical Innovation · Sustainability · Innovation leadership

1 Introduction

The motivation for this research was to explore the success factors for radical innovation within industrial firms in the DACH region (Germany, Austria and Switzerland). This topic has been gaining increasing attention as firms move beyond their more traditional innovation activities (Christensen, 1997; Edwards-Schachter, 2018). This has been accelerated by the need to find new opportunities within highly competitive markets. To support more radical innovation, many firms have used more open collaborations (Chesbrough et al., 2006) to help them innovate outside their traditional incremental approaches. They have used various innovation instruments from start-ups, university collaborations, SME collaborations, innovation labs, and corporate

S. West (✉)
School of Technology and Architecture, Lucerne University of Applied Sciences and Arts, Lucerne, Switzerland
e-mail: shaun.west@hslu.ch

D. Boeni
Zühlke Engineering AG, Zürich, Switzerland

J. Meierhofer et al. (eds.), *Smart Services Summit*, Progress in IS,
https://doi.org/10.1007/978-3-031-36698-7_11

or external venturing. Funding has been from firms, venture funds, and governments (e.g., Horizon).

The authors discovered dissatisfaction with radical innovation from informal discussions with managers of firms leading innovation. Often, innovation was driven by the fear of becoming disrupted or obsolete rather than more constructive drivers. This was in general agreement with the innovation literature, which stated that radical or disruptive innovation is difficult to master and has with it a high chance of failure (Story et al., 2014). This paper by Story et al. (2014) motivates future research, and consequently, the research question for this paper is: "*what are the key factors that impact the likely successes of radical innovation?*".

2 Literature Review

Integrating technology- and non-technology-based innovation is core to a firm's longer-term sustainability (Edwards-Schachter, 2018) and, more widely, further economic growth and social development. Baregheh et al. (2009) state that many different innovation topologies and definitions can confuse us about innovation's meaning or purpose. The existing tools may not measure much of the activity, leading to "hidden innovation" (Miles & Green, 2008). The early theories of Schumpeter (1942) described indicated "heroic" entrepreneurs developing innovation and conversely identified the central innovation agent as large firms. This paradox can be considered at the macro level with the dynamic process of "creative destruction", where old ideas (or structures) become replaced with new ones. To bring this up to date, innovation may be considered a learning process based on multiple interactions (Lundvall, 2013). As such, it is underpinned by multiple relationships between many actors from firms and public and government agencies.

Edwards-Schachter (2018) provides a typology that clarifies innovation into different types: technology, product, process, service, business model, disruptive, radical, design-driven, social, and responsible. Christensen (1997) uses the term "disruptive innovation" to apply a technology that disrupts the market. Innovation that challenges existing business models may be regarded as disruptive, yet it may have little to do with technology. Radical innovation "*sweeps away much of a firm's existing investments in technical skills and knowledge, designs, production technique, plant and equipment*" (Utterback, 1996) as it creates a discontinuous change. Radical innovation moves away from sustaining innovation (maintenance of existing products or services) and incremental innovation. Often radical innovations take time to be developed and gain acceptance in the market, yet when they are accepted, they create a significant. Many such innovations have been based on a shift of "s-curves" in technologies (Sood et al., 2005; Adner et al., 2016). Radical innovations have been created by individual entrepreneurs, small teams, and large firms. Technology-based radical innovation is often led by firms with high R&D intensity (Edwards-Schachter, 2018). Non-technology (e.g., social) radical innovation produces social changes and should be considered as relevant as technology-based radical innovation. Measuring how radical the innovation is can be from either the developer's or adopter's perspective.

Hu et al. (2020) provide insights into radical innovation within family firms to create a research agenda. The insights may be only partially relevant to large firms, but they provide an assessment that can be tested. They break these into four theoretical positions.

i. resource-based view and capabilities;
ii. agency and stewardship;
iii. behavioural agency and socioemotional wealth;
iv. the ability and willingness paradox.

Hu et al. (2020) question the firm's resources and ask if they are the appropriate mix for radical innovation; this includes knowledge and behaviours. Agency and stewardship address the structure of the owner(s) to invest and act on opportunities, yet action may be limited by the alignment of the owning family's goals. Behavioural agency and socioemotional wealth may have some links to the stewardship aspect as they may create risk-averse behaviour, and in this case, it may lead to incremental innovation. The ability and willingness paradox means resources for innovation must be available. The actors must be willing to innovate, driven by the investor's goals and willing to set new goals.

Radical innovation requires direct and indirect human resource management at multiple levels to be successful (Barba-Aragón & Jimènez-Jimènez, 2020). This agrees with the work of Hu et al. (2020), where many of the critical aspects of radical innovation are controlled by people. Their study considered 200 firms in Spain rather than just family-owned firms. Their results concluded that human resources support radical innovation when it is directed towards change, creativity, explorative learning, and competence development. A prior Spanish study (Domínguez-Escrig et al., 2019) confirmed the importance of "stewardship behaviour" in mediating radical innovation. Stewardship is where human resource management can support a firm's development. Stewardship theory states that managers are motivated by individual goals and collective/pro-organization goals for the common good (both short- and long-term. The study confirmed that firms whose leaders showed high levels of stewardship created an impact in society, addressed social issues and global threats and, in doing so, created a favourable climate to foster (radical) innovation. These impacts can be understood within the sustainability context of the triple bottom line (Slaper et al., 2011), which is an accounting framework comprising social, environmental and economic dimensions.

3 Methodology

A qualitative research framework would be appropriate for this study based on the research question posed in the introduction. A literature review was considered. However, interviews were more appropriate given the authors' direct access to senior management in innovating firms. This approach is supported by Bell et al. (2022). Selection and individual bias were removed through a section of firms primarily in the DACH region of Europe, ranging from start-ups and SMEs to large multinationals. The target group of businesses was industrial firms; within these, the aim was to remove bias by interviewing at different levels and functions. The authors considered the choice of a survey, semi-structured interviews, or mixed methods. Semi-structured interviews were considered the best approach for an initial study to understand the success factors for radical innovation and are confirmed by Bell et al. (2022) as appropriate for an exploratory study. To verify the results and refine them, a two-round Delphi process (Linstone et al., 1975) was used to ensure the consistency and validation of the data collated, and the insights gained.

The initial contact list for the interviews was based on the firms' CRM system, the authors' relevant contacts and LinkedIn. Two hundred contacts yielded 100 interviewees. The interviews took place in two tranches, from March to July 2021 for the initial interviews and from July 2021 to September 2021 for the follow-up reviews/confirmations. All the interviews were recorded where permission was given; otherwise, interview notes were taken and agreed upon with the interviewee. Recordings were made, transcripts were created, insights extracted, and representative quotes were captured from the transcripts. Then, over three months, the interview themes were clustered, reflected upon and re-clustered in an irritative process. Finally, an anonymised report was developed, and the finding presented to the interviewees (80% return rate) for confirmation and verification.

4 Results

The results consider first the barriers to radical innovation and then lessons from the interviews that can make radical innovation more sustainable in industry.

4.1 Barriers to Radical Innovation

The Delphi identified seven barriers to radical innovation within firms in the DACH (Table 1). Reducing the insights from the interviews took several iterations to reduce them into the seven barriers. In each case, the agreed cause based on the two-cycle Delphi is presented in Table 1 with a representative quote to provide contextual insight into the root cause. The confirmed cause is described in the table, and this explorative study excludes the extreme cases or analysis based on business size etc.

Table 1 Seven barriers and their cause

Barriers	Central root cause and representative quote
Decision-makers and processes	Decision-makers do not understand and have no experience of radical innovations. False expectations, KPIs and inappropriate empirical values stand in the way of the projects *"Competent top managers from established companies are usually unqualified as decision-makers for radical innovations"*
Operationalization and scaling	Once a radical innovation has made it to market, it is difficult to put it into operation and to market it on a large scale. Scaling requires an organization to grow and evolve as it transitions from development to delivery *"We clearly underestimated the scaling after entering the market, it was extremely intensive and painful"*
Sales	The existing sales department in most cases is neither able nor incentivized to successfully sell the radical innovation. At the same time, no new distribution may be set up *"Over 90% consider their existing distribution unsuitable to sell the new product or service"*
Product management/ product market fit	Solutions for which there is no market are prioritized and promoted. Finding product-market-fit is difficult *"Product solutions are pushed forward with a lot of zeal, even if there is no market for them"*
Speed	Almost all the interviewees complained about insufficient speed, caused by rigid processes, a lack of decisions, a lack of error culture and a lack of attractiveness on the employee market *"We're just way too slow!"*
Budget and investment	The majority of radical innovation projects do not have their own budget and are unable to access sufficient growth investments comparable to start-ups *"While start-ups go full throttle, I have to have a formal application approved for every new laptop"*
Conflicts of interest	The "owner" of the radical innovation is often caught between the operational business and the owners of the firm. This leads to conflicts *"Every executive board meeting I had the same fights and discussion with the CFO and the CSO about ROI and short-term versus long-term"*

4.2 *Lessons to Make Radical Innovation More Sustainable*

The reduction of insights from the interviews took several iterations to produce ten lessons (Table 2). Again, the results are the outcome of a two-cycle Delphi and are presented in the table with a representative quote. The lessons here are the most common and are written generically. Again, as this is an explorative study, the table excludes extreme cases or analysis based on business size etc.

Table 2 Lessons that help improve the sustainability of radical innovation

Issue	Lesson and representative quote
Passionate drive	It takes a person with unrelenting drive, the necessary resilience and tenacity. They need to take the lead to push through the initiative for years to come *"In all successful initiatives we have seen a clear driver who rallied the team behind them and spearheaded necessary internal conflicts. If this person was missing the initiatives didn't make it beyond the first serious issues"*
Personal responsibility	The team must take and be allowed to take full responsibility for the initiative: from development to product definition to sales and marketing *"While start-ups speed on we lose weeks for every minor investment decision"*
Time	Radical initiatives require about 10 years of investment before they turn a profit. This applies to start-ups as well as to corporate initiatives *"When I harvest the crops before they are ripe, I lose the entire harvest"*
Decoupling	Radical innovations are meant to create something new. The new therefore also requires its own culture, attitude, KPIs, etc.–and a corresponding decoupling from the existing company *"We continued to separate more and more over time. In hindsight, we should have separated much stronger from the beginning"*
Think holistically	To prevent tunnel vision, diverse/external perspectives help, especially for the product definition and scaling of the market presence *"We discussed detailed technical implementations while nobody believed that our existing sales team can actually sell it. Still, nobody builds a suitable sales team"*
Approach/ Methodology	We have not seen a single project that had problems finding the right methodology for the current needs
Vision and strategy	Less than 3 percent of all interviews mentioned corporate vision and strategy *"Innovation initiatives triggered by an overarching vision of the established company led to too big, micromanaged teams which consisted of the wrong people"*
Technology	Teams often focused on technology, although it rarely became a critical factor in radical innovation *"We put a lot of effort in proving the technology early even though we later realized that the challenge comes from the market and the proof of technology has to be redone anyway due to changing requirements"*
Trends	The study showed no successful innovation initiatives triggered by trend scouting or similar trend led activities and focus. The result was rather the opposite, where trend triggered initiatives never made into a first-generation product but stopped at the proof of concept. *"A great initiative not following a megatrend is still a great initiative A megatrend without a suitable team and idea is still nothing"*
Customer/ Investor	In established companies the investment structure for internal innovation initiatives is mostly unclear. This leads to conflicts in timeline and feature creep, as the internal customer continuously requests new features without confirming their need *"Our internal departments all continuously want more features, blowing up our products. But as we rely on their investments, we can hardly deny them, leading to a bloated and complex product"*

5 Discussion

Our interviews paint a clear picture: radical innovation happens where the owners have the understanding and long-term focus to allow and sustainably promote radical innovations. This happens when a board-level member with the necessary experience and qualifications implements this mandate from the owners and supports the project over the entire period, and also, where a single person has the strength to implement a large, long-term vision piece-by-piece with a motivated and self-critical team. These building blocks are mandatory prerequisites for success, but you also need a team with the right qualities. These include supposedly soft, but success-critical factors such as diligence, talent, intelligence, team spirit, culture, willingness to deal with conflict, know-how, technology, etc. and—most notably, —a high degree of tenacity.

Unfortunately, there is no universal recipe for success in radical innovation. Radical innovation is a highly complex, highly individual challenge for which no universal model solution exists. Each project and each situation must be considered individually and transparently. It requires observation, hypotheses, and testing, followed by reflection and adjustment of points of view based upon new information. However, the discussions show that specific frameworks and methods can help to make the process more efficient if they are correctly understood and used in their context.

During this study, based on active consulting mandates and personal experience from our start-ups and radical innovation projects, we have developed a framework with which we support people, teams, and managers. It supports them in structuring their projects more systematically, communicating more efficiently and focusing on critical issues.

This paper has provided insights sufficient to identify seven barriers to radical innovation and simultaneously identified the actions that can support radical innovation. Nevertheless, this is an explorative study and demands further analysis to improve the definition of the barriers and has only started to outline what 'best practice' could be. It helps the firm's sustainability because firms need to evolve and innovate to face new challenges. Radical innovation is necessary for the long-term survival of the firm. It is also needed when considering 'people, plant and profit' or the UN's Sustainable Development Goals, where to achieve some of these goals, radical innovation rather than incremental innovation is needed.

5.1 Managerial Relevance

Radical innovation is people-driven, and it requires an open-minded learning firm to achieve this. Radical innovation needs management to understand what it is and to provide it with the necessary resources, as it requires more than just money, and management has to understand the timeframe. It is also necessary to appreciate that radical means constructive destruction—in the longer term, you may destroy your existing business and replace it with something new.

The execution of radical innovation needs stewardship and guidance and, at times, protection from the traditional business. Traditional top-down management (i.e., CEOs) are taught to lead a business based on incremental improvements, so traditional leadership needs to move to stewardship models to support radical innovation. This also is relevant for the business owner who needs to understand that the radically innovative part of the business is operationally different to the 'normal' business.

5.2 *Academic Implications*

The literature has been quite explicit with the term radical innovation, yet there is a great deal of uncertainty about its actual meaning in industry. This may be because of the narrower views that managers in individual firms have, or it may be that the firms' vision is limited. Alternatively, it may be a "hidden innovation" (Miles & Green, 2008) that is ongoing and unknown to the interviewees. This nevertheless needs investigation as our innovation opportunities, as described by Edwards-Schachter (2018), are limited and remain trapped and unable to create a future without breaking the present (Schumpeter, 1942).

The barriers to radical innovation that we identified could all be linked directly or indirectly to the human aspects described by Barba-Aragon & Jimenez-Jimenez (2020). This could be at the ownership/investor level, C-level, or middle management level. Leaders could support innovation through stewardship concepts, which need to be maintained to support creativity, explorative learning, and competence development necessary for radical innovation. The day-to-day aspects create a culture where operational efficiency remains more imperative than innovation, even though the theory tells us that we need innovation to be sustainable in the long run.

6 Conclusions and Future Research

To become effective at radical innovation, the firm must become a learning firm and one that considers aspects outside the business. The firm needs to consider the behavioural aspects of management, the necessary adaptability of the process and the long-term funding aspect of radical innovation. This way, the firm can be sustainable and support the achievement of the UN's Sustainable Development Goals.

As an exploratory study, more work is needed to assess the data quantitively. This should include ranking the barriers and further integrating them with actions to overcome them.

References

Adner, R., & Kapoor, R. (2016). Innovation ecosystems and the pace of substitution: Re-examining technology S-curves. *Strategic Management Journal, 37*(4), 625–648.

Barba-Aragón, M. I., & Jiménez-Jiménez, D. (2020). HRM and radical innovation: A dual approach with exploration as a mediator. *European Management Journal, 38*(5), 791–803.

Baregheh, A., Rowley, J., & Sambrook, S. (2009). Towards a multidisciplinary definition of innovation. *Management Decision.*

Bell, E., Bryman, A., & Harley, B. (2022). *Business research methods.* Oxford University Press.

Chesbrough, H., Vanhaverbeke, W., & West, J. (2006). Open innovation. *The new Imperative for creating and profiting from technology, 1.*

Christensen, C. (1997). Patterns in the evolution of product competition. *European Management Journal, 15*(2), 117–127.

Domínguez-Escrig, E., Mallén-Broch, F. F., Lapiedra-Alcamí, R., & Chiva-Gómez, R. (2019). The influence of leaders' stewardship behavior on innovation success: The mediating effect of radical innovation. *Journal of Business Ethics, 159*(3), 849–862.

Edwards-Schachter, M. (2018). The nature and variety of innovation. *International Journal of Innovation Studies, 2*(2), 65–79.

Hu, Q., & Hughes, M. (2020). Radical innovation in family firms: A systematic analysis and research agenda. *International Journal of Entrepreneurial Behavior & Research, 26*(6), 11991234.

Linstone, H. A., & Turoff, M. (Eds.). (1975). *The delphi method* (pp. 3–12). Addison-Wesley.

Lundvall, B. Å. (2013). Innovation studies: A personal interpretation of the state of the art. *Innovation Studies: Evolution and Future Challenges, 21.*

Miles, I., & Green, L. (2008). *Hidden innovation in the creative industries*

Schumpeter, J. (1942). Creative destruction. *Capitalism, Socialism and Democracy, 825*, 82–85.

Slaper, T. F., & Hall, T. J. (2011). The triple bottom line: What is it and how does it work. *Indiana Business Review, 86*(1), 4–8.

Sood, A., & Tellis, G. J. (2005). Technological evolution and radical innovation. *Journal of Marketing, 69*(3), 152–168.

Story, V. M., Daniels, K., Zolkiewski, J., & Dainty, A. R. (2014). The barriers and consequences of radical innovations: Introduction to the issue. *Industrial Marketing Management, 43*(8), 1271–1277.

Utterback, J. M. (1996). *Mastering the dynamics of innovation.* Harvard Business School Press.

Empowering Sustainability and Innovation Session

SSDL: A Domain-Specific Modeling Language for Smart City Services

Rubén Ruiz-Torrubiano, Deepak Dhungana, Gerhard Kormann-Hainzl, and Sarita Paudel

Abstract As software services become more and more indispensable in our daily lives, low-code and no-code platforms are gaining significance, especially for non-technical users. We present a novel domain-specific modeling language for the definition of smart service systems in the context of smart cities, which we call Smart Service Definition Language (SSDL). The main goal of this formal language is to provide a basis for a low-coding software engineering approach for the design and deployment of smart services. SSDL is based on the Smart City Ontology (SCO) to provide syntactic and semantic elements related to the smart city environment and aims at defining a syntax as near as possible to human language to overcome acceptance problems for non-technical users. The proposed language can be used by smart system designers and other stakeholders to define the components, actors, data and relationships between the different elements that compose smart service systems in a formal and reproducible way, paving the way for an automatic or semi-automatic generation of ready-to-deploy smart services as independent applications, which may take the form of web services to be integrated in service-oriented architectures. In this paper, we present the syntax of SSDL and provide an example use-case for its application to the problem of designing a smart service for parking lot management.

R. Ruiz-Torrubiano (✉) · D. Dhungana · G. Kormann-Hainzl · S. Paudel
IMC University of Applied Sciences Krems, 3500 Krems, Austria
e-mail: ruben.ruiz@fh-krems.ac.at

D. Dhungana
e-mail: deepak.dhungana@fh-krems.ac.at

G. Kormann-Hainzl
e-mail: gerhard.kormann@fh-krems.ac.at

S. Paudel
e-mail: sarita.paudel@fh-krems.ac.at

J. Meierhofer et al. (eds.), *Smart Services Summit*, Progress in IS,
https://doi.org/10.1007/978-3-031-36698-7_12

1 Introduction

Service systems can be defined as dynamic value-cocreation configurations of resources, connected internally and externally to other service systems by value propositions (Maglio et al., 2009). This configuration includes in general people, organizations, shared information and technology. Service consumers and providers interact by creating value in the form of a value proposition, which is an invitation from actors to one another to engage in service (Chandler & Lusch, 2015). The combination of these concepts with advances and developments in information technology (IT) results in the emergence of smart service systems (SSS), which are software systems capable of learning, dynamic adaptation, and decision making based upon received and transmitted data (Lim & Maglio, 2018). In general, smart service systems are capable of monitoring, optimizing and controlling smart products and devices to deliver value for the service participants. In the smart city context, smart service systems enable value-cocreation for the participants of the smart city ecosystem to address its main challenges (like urbanization, climate change, sustainable transport, housing, and healthcare) by an intelligent use of information technologies (Wolff et al., 2020). One of the main challenges to deliver these goals is an efficient way of engineering smart service systems to enable the city and its citizens to harness the large volumes of data that are produced by sensors and digital infrastructure to improve sustanability by service innovation (Dobler et al., 2021; Suzic et al., 2022). In this paper, we address this software engineering challenge by proposing a low-coding approach based on a domain-specific modeling language that we call Smart Service Definition Language (SSDL). The basic goal of this language is to provide a means of defining, sharing and deploying smart service systems in a natural way also for non-technical users, enabling the relevant stakeholders in the city administration to autonomously define and deploy smart service systems based on their needs and the available infrastructure. SSDL is the foundation for a low-coding platform capable of generating applications based on a simple formal description, largely simplifying the development of smart services in practice.

This paper is organized as follows. In Sect. 2 we begin by providing a review of related work in the field of smart systems design with a focus on smart cities. Section 3 outlines the research and design methodology used and provides the syntax and semantics definitions for SSDL. In Sect. 4 we describe a general software architecture for implementing a low-coding platform for smart service systems based on SSDL. Our approach is evaluated in Sect. 5 by applying it to an example use case. We conclude in Sect. 6 with a summary.

2 Related work

Smart products integrate resources and activities of service providers and service consumers, which allows them to adapt the service systems based on contextual data (Beverungen et al., 2019a). In Beverungen et al. (2019b), the authors present the concept of smart services and smart service systems. As smart products are widely used for smart sevices in smart cities the related work helps us better understand our target domain. Jussen et al. (2019) present a service-engineering approach for smart services and illustrates its successful application and its impact on a medium-sized company. The study focuses on the development steps involved and the interaction and interconnection of elements in smart services. Komninos et al. (2016) propose an ontology for smart cities called Smart City Ontology (SCO) and present its building blocks (e.g., technology, structure, functions, design), and properties for connecting those blocks in the ontology. This study was enhanced in Komninos et al. (2021) with an OWL ontology. In this work, we use this OWL-based Smart City Ontology to provide the language with semantic and syntactic features related to smart cities, therefore these related works help us to understand the smart city landscape, identify its main components and the processes involved in smart city applications. In Huber et al. (2019) the authors develop a domain-specific modeling language for smart service systems and demonstrate it by presenting real-world scenarios. This work focuses on concepts of domain-specific modeling and relationships in smart service system domains, highlighting the importance of developing domain-specific modeling languages for better coverage of the target domain. Similarly, guidelines for selecting an appropriate metamodeling language are presented in Frank (2013). The authors demonstrate a process for specifying a domain-specfic modeling language based on requirements analysis. We considered this work as our starting point for designing our domain-specific modeling language.

3 Smart Service Definition Language

SSDL is a text-based language. An SSDL file contains the definition of a smart service. For the general structure, we borrowed some syntactic elements from YAML,[1] as this language provides syntactic constructions that can be regarded as near to natural language. We then enriched these constructions by defining additional semantic elements from the Smart City Ontology. We followed the general methodology for designing domain-specific modeling languages proposed in Frank (2013). This methodology comprises the following steps: (1) clarification of scope and purpose, (2) analysis of generic and specific requirements, (3) language specification, (4) design of graphical notation, (5) development of a modeling tool, and (6) evaluation and refinement.

Table 1 shows a summary of the syntax of SSDL. The preamble contains metadata, i.e. general data about the smart service itself, mainly for documentation purposes. Two metadata-fields are mandatory: `name` and `version`. Additional metadata-fields can be used to define the general area that the smart service belongs to, like environment or transportation.

[1] https://yaml.org/.

Table 1 Structure of the Smart Service Definition Language (SSDL)

Section	Description	Elements	Example
Service	Metadata, documentation, classification	`name`	`name is "Waste Management"`
		`version`	`version is 1.0`
		`domain`	`domain is Infrastucture`
		`field`	`field is Environment`
Data	Configuration data sources, sensors, gateways, formats	`name`	`name is "Waste Bins"`
		`type`	`type is WasteSensor`
		`provider`	`provider is Fiware`

		`config`	`config:`
			`url is http:...`
			`token is 0309f94...`

		`query`	`query:`
			`sensor: urn:...:123`
			`property: capacityRemaining`
			`property: location`
			`property: time`

		`format`	`format:`
			`capacityRemaining is number`
			`location is geodata`
			`time is timestamp`

Application	Service definition and data visualization	`type`	`type is WebApplication`
		`layout`	`layout is SinglePage`
		`roles`	`roles:` `administrator, user`

		`visualization`	`visualization:`
			`type is LineChart`
			`x is time`
			`y is capacityRemaining`

Deployment	Deployment profiles	`type`	`type is Docker-Compose`
		`file`	`file: docker-compose.yaml`

		`credentials`	`credentials:`
			`username: admin@smart.city`
			`password: 123adjikj!...`

The `data` section contains definitions regarding the data sources used, which typically belong to the sensor infrastructure already deployed in the smart city. In this section, one or more data sources can be defined using a YAML-like list notation. The data sources themselves can be specified using the following properties: the `name` of the data source, the `type` of the data source used and the name of the Internet-of-Things (IoT) `provider` (e.g. Fiware[2]). Compound objects are used to specify additional details: `format` specifies the data returned by the data source, including property names and types, `config` describes how to authenticate against the IoT platform, and `queries` defines how to query specific sensors or entities.

The `application` section contains the declarations needed for specifying the smart service itself. This includes properties like the type of the application and how to display data. The properties that can be specified are: the `type` of the application, which will be a web application by default, the `layout` (like single-page or other types of layouts), the standard user `roles`, like administrator and a read-only user type, and the data `visualization` used, like several types of plots or tables.

The `deployment` section is optional and specifies deployment environments where the smart service will be published and made available to the end users. A deployment environment is a runtime or framework being executed on the server that runs the smart service system. By defining the runtime environment in the SSDL file, the user can use tools for automating the deployment process and thus decrease time-to-market as much as possible.

4 Architecture

In this section, we outline a general software architecture for a low-coding platform based on SSDL. An overview of this architecture is given in Fig. 1. We begin with the user interface (UI) component at the bottom, which enables the user to interact with the system. The central element of the UI is a text editor featuring syntax highlighting (rendering language keywords and symbols differently) and validation (providing visual cues like underlining incorrect syntax and explanations in text form).

Low-coding features like pre-defined blocks of code that can be dragged directly into the editor are also integrated in the UI. For instance the user can choose from a sidebar between different pre-defined components like default data blocks (data section), default application (web applications) and deployment types (container-based deployment). The configuration of these blocks is done visually by means of configuration dialogs that can be used to save the configuration or modify it later.

The backend is organized as a service-oriented architecture (SOA) composed of several independent services, each providing a subset of the needed functionality. The main services composing the backend are the `ApplicationService`, the `LanguageService` and the `DataService`. The `ApplicationService` is the central business logic service for the web application that communicates with

[2] https://www.fiware.org/.

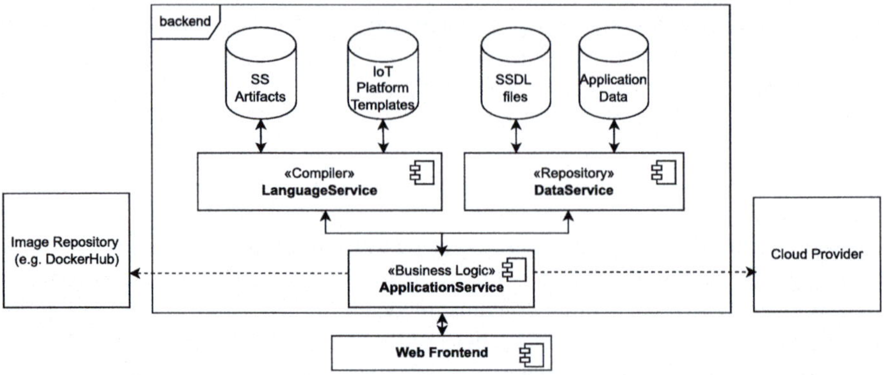

Fig. 1 An architectural overview for a low-code platform based on SSDL

the frontend to provide the requested functionality. Its main task is to serve as a controller for the features supported by the UI and a proxy to the other services by exposing the necessary APIs. The `LanguageService` implements a compiler for SSDL and generates the source code of the corresponding smart service (which we call the *artifact*). This service is also responsible for handling validation requests from the UI, which are forwarded by the `ApplicationService`. The output of the `LanguageService` is a fully fledged web application based on a general programming language (like Java or Python). The compiler generates the code of the application based on *templates*, which are in essence source files with placeholders that are filled at compile time. These templates can be either general (concerned with the general structure of the application and the main business logic) or IoT-specific, providing modules implementing API clients for the target IoT platform. For instance, if the sensor layer is based on the Fiware framework, the template that implements the Fiware API is selected by the compiler. The implementation is assembled in compile time with the smart service so that the service, when deployed, can access and fetch data from the corresponding sensor layer. This encapsulates the concrete data modeling for a specific IoT vendor and makes SSDL independent of the specific IoT platform used. Finally, the `DataService` is responsible for managing a repository of source SSDL files. These files are kept in a data store. Additionally, this services manages all the other types of data needed by the application, like user data.

5 Evaluation

We now present a use-case solved using the proposed domain-specific language in the smart city context: smart parking. The problem of managing the available parking space in modern cities has practical importance for both the drivers and the operators and has sustanability implications. There are several specific variants of this problem depending on the goals and the technology used (Al-Turjman & Malekloo, 2019). In general, smart parking services alleviate congestion in city centres and thus reduce vacant slot search time which also results in reduced air pollution. The sensing layer consists in sensors placed in the available parking space so that the presence or absence of a vehicle can be measured. The sensors then transmit measurements using the corresponding network infrastructure to the IoT middleware. In a traditional smart service engineering approach, dedicated application development would be needed to build a tightly coupled application that communicates with the IoT middleware to fetch data and implement some functionality related to the data. In contrast to this approach, we now build an SSDL smart service definition in Fig. 2. In this definition, sensors of infrared type and four data fields of interest are specified: an identifier for the sensor making the measurement, the timestamp of the measurement, a boolean indicating if the parking slot is vacant or not, and the location of the parking slot. When compiled, this section would generate an implementation of the NGSIv2 REST API.[3] Next, the application is defined as a single-page web application with standard administrator and user roles. The application shows the data in two ways: first, it displays a map showing the city centre within a radius of 10 km where the data points are defined by their GPS coordinates and the boolean indicating if the spot is vacant or not. Additionally, it shows the data as a table. Note that all that we need is to specify a few lines of text instead of writing a full web application from scratch. This lowers the costs of developing such an application and at the same time serves as an implementation guide for further smart cities interested in introducing smart parking solutions themselves.

6 Conclusions and Future Work

In this paper, we presented a new domain-specific modeling language and a general software architecture for implementing a low-coding approach for developing smart services in the smart city context. The main contributions of the present study are the following: first, a general and flexible declarative formal language for automatically generating smart service applications was proposed. This language is designed by using a principled methodology and resembles natural language to booster acceptance by non-technical users. Second, SSDL represents the foundation for a low-coding platform for engineering smart services, allowing for implementing a wide variety of features for supporting the user (e.g. graphical modeling tools). Third, our approach is IoT platform agnostic, which means that it can work with arbitrary data models and middlewares. Adaptation to new platforms can be done by implementing an abstract layer that supports the corresponding middleware API.

Current ongoing work includes the implementation of a platform prototype to demonstrate its feasibility. Visualization features (like representing code blocks to be

[3] https://fiware-orion.readthedocs.io/en/latest/.

```
service:
  name is "smart_parking"
  version is 1.0
  domain is Infrastructure
  field is Transportation
data:
  - sensor1:
    provider is Fiware
    config:
      endpoint: http://...
      token: 0af2347ed...
    name is "parking_lot"
    type is Infrared
    queries:
      entity: urn:parking:434
      property: id
      property: time
        timestamp
      property: vacant
        bool
      property: location
        geodata
application:
  type is Application.Web
  roles:
    administrator, user
  visualization:
    - "map1":
      type is Map
      center is 48.21,16.36
      data:
        x is location
        y is vacant"
    - "table1":
      type is Table
      data:
        - id, timestamp,
        location, vacant
deployment:
  type: Docker-Compose
  file: "docker-compose.yaml"
```

Fig. 2 Example SSDL file for a smart parking service

used graphically, and using drag-and-drop for connecting these components) should be investigated to determine their viability as well. The prototype will be further developed to implement an example use case from end to end, providing a sample implementation for first production usages.

References

Al-Turjman, F., & Malekloo, A. (2019). Smart parking in IoT-enabled cities: A survey. *Sustainable Cities and Society, 49*, 101608.

Beverungen, D., Breidbach, C. F., Poeppelbuss, J., & Tuunainen, V. K. (2019a). Smart service systems: An interdisciplinary perspective. *Information Systems Journal*, *29*(6), 1201–1206. Publisher: Blackwell Publishing Ltd.

Beverungen, D., Müller, O., Matzner, M., Mendling, J., & vom Brocke, J. (2019b). Conceptualizing smart service systems. *Electronic Markets*, *29*(1), 7–18. Publisher: Springer Verlag.

Chandler, J. D., & Lusch, R. F. (2015). Service systems: A broadened framework and research agenda on value propositions, engagement, and service experience. *Journal of Service Research*, *18*(1), 6–22. Publisher: SAGE Publications Inc.

Dobler, M., Kalkhofer, H., & Schumacher, J. (2021). Smart service development in public-private settings–assessment methodology and use-cases in the lake constance region. In S. West, J. Meierhofer, & C. Ganz (Eds.), *Smart services summit* (pp. 3–13). Cham. Springer International Publishing.

Frank, U. (2013). Domain-specific modeling languages: requirements analysis and design guidelines. In *Domain engineering: product lines, conceptual models, and languages* (pp. 133–157). Journal Abbreviation: Domain Engineering: Product Lines, Conceptual Models, and Languages.

Huber, R. X. R., Püschel, L. C., & Röglinger, M. (2019). Capturing smart service systems: Development of a domain-specific modelling language. *Information Systems Journal,29*(6), 1207–1255.

Jussen, P., Kuntz, J., Senderek, R., & Moser, B. (2019). *Smart service engineering* (Vol. 83, pp. 384–388). Elsevier B.V.

Komninos, N., Bratsas, C., Kakderi, C., & Tsarchopoulos, P. (2016). Smart city ontologies: Improving the effectiveness of smart city applications. *Journal of Smart Cities, 1*(1).

Komninos, N., Panori, A., & Kakderi, C. (2021). *The smart city ontology 2.0: Assessing the components and interdependencies of city smartness*. Publisher: Preprints.

Lim, C., & Maglio, P. P. (2018). Data-driven understanding of smart service systems through text mining. *Service Science, 10*(2), 154–180. Publisher: INFORMS.

Maglio, P. P., Vargo, S. L., Caswell, N., & Spohrer, J. (2009). The service system is the basic abstraction of service science. *Information Systems and e-Business Management, 7*(4), 395–406.

Suzic, B., Urban, S., Hellwig, M., & Dobler, M. (2022). Smart circular economy value drivers: The role of the financial sector in stimulating smart regional innovation-driven growth. In S. West, J. Meierhofer, & U. Mangla (Eds.), *Smart services summit* (pp. 55–64). Springer International Publishing.

Wolff, A., Barker, M., Hudson, L., & Seffah, A. (2020). Supporting smart citizens: Design templates for co-designing data-intensive technologies. *Cities, 101,* 102695.

Target Group Generation Z—How Stationary Retailers Can Fulfil the Expectations of Young Customers Through Smart Services and Technologies to Increase Their Resilience

Stefan Schweiger, Julian Kemmer, Simon Oed, Julius Taubert, Valerie Bass, Michael Eddelbuettel, and Tamara Wuerstle

Abstract This paper aims to apply the basics of the Service-Dominant Logic, especially the concept of creating benefits through serving, to the stationary retail industry. In the industrial context, the shift from a product-driven point of view to a service-driven perspective has been discussed widely. However, there are only few connections to how this can be applied to the retail sector on a B2C-level and how retailers can use smart services in order to enable customer engagement, loyalty and retention. The expectations of customers towards future stationary retail develop significantly as consumers got used to the comfort of online shopping. Especially the younger generation—the Generation Z—seems to have changed their priorities from the bare purchase of products to an experience- and service-driven approach when shopping over-the-counter. To stay successful long-term, companies from this sector need to adapt to the expectations of their future main customer group. Therefore, this paper will analyse the specific needs of Generation Z, explain how smart services contribute to creating benefit for this customer group and how this affects the economic sustainability of these firms.

Keywords Stationary retail · Generation Z · Digital services · Economic sustainability · Customer journey

S. Schweiger (✉) · J. Kemmer · S. Oed · J. Taubert · V. Bass · M. Eddelbuettel · T. Wuerstle
HTWG Konstanz—University of Applied Sciences, Alfred-Wachtel-Str. 8, 78462 Konstanz, Germany
e-mail: smart-service-bw@htwg-konstanz.de

J. Meierhofer et al. (eds.), *Smart Services Summit*, Progress in IS,
https://doi.org/10.1007/978-3-031-36698-7_13

1 Introduction

In the past 15 years, one of four German stationary retailers has had to shut down their businesses (HDE 2022). The COVID-19 pandemic and the war in Ukraine which resulted in rising inflation and supply shortages are contributing to a further intensification of the situation. However, the precarious situation of many retailers is not only due to these factors. Other reasons for the existential threat to many retailers are digitalization and growing online competition with partly disruptive business models (Batsakis et al., 2022). In order to be able to ensure long-term competitiveness in terms of economic sustainability, the retail sector, which is considered to be weak in innovation, should be aware of consumer demands and address the requirements of all generations, in particular of young digital natives who have been neglected for a long time. Especially the demands and expectations of Generation Z pose a major challenge for retailers (Heinemann, 2021). However, digital services and technologies offer numerous new digital touchpoints and opportunities for retailers to interact with this young and important group of consumers along all phases of their customer journey. This paper aims to evaluate these possibilities based on consumer expectations in order to preserve the economic sustainability of the stationary retail sector and to increase its resilience.

2 Literature Review

Services and Smart Services in the Service-Dominant-Logic

In this paper, the lens of the concept of Service-Dominant-Logic (in this paper referred to as S-D logic) has been applied. Vargo and Lusch describe the S-D logic as the focus on the process of serving instead of the actual output namely a specific product offering (Vargo & Lusch, 2016). According to them, S-D logic considers the benefit that a company creates for another actor through the exchange of services as more relevant than the offering of output which can be tangible or intangible goods. Within the S-D logic, the term service is described as the process of creating benefit for specific actors. Whereas previous definitions understood the term service as the output of immaterial goods. This changes the concept and definition of the term service. Even though in the industrial context, this point of view has been widely analysed in the past years, this concept has barely been applied to the retail sector. However, services and especially smart services hold significant potential for retailers to increase customer benefit, engagement and loyalty and hence increase their long-term success in terms of economic sustainability.

Customer Journey

From first contact to purchase—the term customer journey encompasses the entire experience of a potential customer, during which he encounters a brand, a service, or a product through various touchpoints, up to the point of purchase and beyond

(Naskrent et al., 2021). It is crucial that the customer journey does not end with the point-of-purchase. Instead, the post-purchase phase must be considered in addition to the pre-purchase and purchase phases (Esch et al., 2010). After the customer's attention is drawn to the product by an impulse in the pre-purchase phase, an attempt is made to convince him of the product with the help of a wide variety of touchpoints. A touchpoint is understood to be the possible interfaces and points of contact between a customer and a company. These can be digital touchpoints (apps, website, social media) as well as traditional touchpoints (catalogs and trade shows) (Wang et al., 2022). After the customer has been convinced by the product, the second phase begins, in which the product is purchased. Subsequently, the aim of the post-purchase phase is to continue to support the customer as effectively as possible after the purchase and to underpin customer loyalty (Sivapalan et al., 2022).

Generation Z

According to the historical-social terminology, a generation is a group of individuals who share the same period of birth as well as formative events in childhood and adolescence. The resulting values, attitudes and views of the generation thus differ between generations (Klaffke, 2021). Generation Z discussed in this paper was born between 1995 and 2009 and follows Generation Y (1980–1994), which in turn came after Generation X (1965–1979) and the Baby Boomer Generation (1950 and 1964) (Maas, 2022).

Generation Z was raised in a globally connected world and grew up with the internet and other digital technologies. Although Generation Z is considered to be extremely affine to the internet, they also take conscious time-outs from the online world regularly (Merkle, 2020). Generation Z is looking for recognizable individuality, which can also be expressed in social networks and thus generates appreciation and attention. In the field of consumption, representatives of Generation Z are considered demanding, well-informed and highly influential. As a result, this age cohort often has a great influence on the purchasing decisions of their parents and grandparents, which illustrates an additional relevance for stationary retailers (Kleinjohann & Reinecke, 2020).

3 Methodology

In order to gain insights into Generation Z's expectations of stationary retail, data was collected in an empirical study conducted by the research project "Handel innovativ" at the HTWG Konstanz—University of Applied Sciences. In the period from May to June 2022, a total of 178 members of this generation from German-speaking countries completed an anonymous online survey. The participants were asked about their requirements and expectations of the over-the-counter retail sector using a standardized questionnaire. The questions of the online survey were designed to capture both consumer behavior and the requirements of stationary retail in all phases of the customer journey. Although the survey was primarily aimed at Generation Z, people of all ages were able to participate in the survey. In addition, some data sets from older generations (53 from Generation Y, 47 from Generation X and 38 from Baby Boomers and older) were obtained and used for comparison purposes and to identify trends between generations in this study.

4 Results

The results of the survey indicate that young consumers demand both online and stationary retail touchpoints. A recognisable preference for online and offline retail does not exist according to the data. Thus, Generation Z expects digital and analogue touchpoints to retailers along all phases of the customer journey and a cross-channel shopping experience. This requires interconnected sales and communication channels within the framework of a cross- or omni-channel concept, which accompanies the consumer along the purchasing process and enables changing of channels at any time along the process (Thaichon et al., 2022). The relevance of connected shopping across multiple channels is increasing from generation to generation. Only 11% of the Baby Boomers rate cross-channel shopping as important or very important, whereas this number rises to 41% among Generation Z. However, clear preferences emerge in the retail channels demanded by young Generation Z consumers. According to the survey, a high proportion of respondents from all generations demand a website. Especially for young consumers of generations Y and Z, an online shop from retailers who primarily operate in stationary shops is also important. More than one third of Generation Z representatives also want a social media presence. This percentage is significantly lower among the digital immigrants of Generation X and the Baby-Boomers. It is also notable that Baby-Boomers and older participants in particular are considerably more likely to attach less value to digital customer touchpoints than digital natives from Generation Z and Y do. In addition to interconnected channels, young consumers also demand digital services and technologies. These should make over-the-counter shopping easier, more convenient and in some cases more entertaining and hence create benefit in the sense of the S-D logic (Vargo & Lusch, 2016). This request is also significantly more noticeable among young consumers than among older participants in the survey. However, a more detailed assessment revealed that specific digital services and technologies are expected in particular. Figure 1 shows the proportionate expectation of selected digital services and technologies in the retail sector. The survey participants were able to select several of the given options.

In general, it can be seen that the expectation of digital services and technologies is increasing from generation to generation. The data collected show that supporters of all generations demand cashless payment via debit or credit card. However, payment via smartphone and smartwatch is also becoming increasingly important, especially among young consumers. Transparent inventories, which can be viewed in real time on any channel, are also extremely important, especially among members of Generation Z. Extended digital product information, loyalty and bonus programs, as well

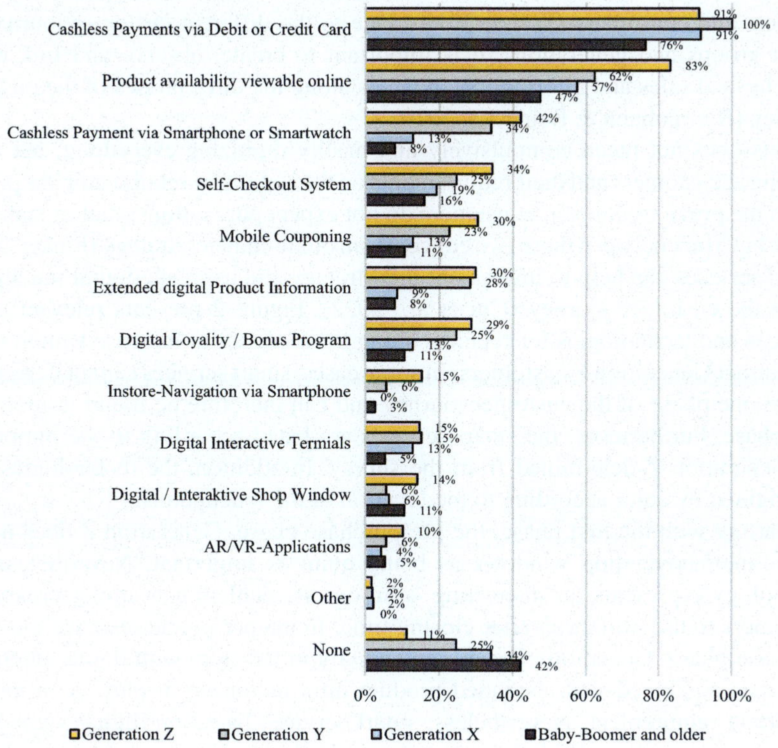

Fig. 1 Demanded digital services and technologies

as personalized digital offers via app are also popular among young consumers. Self-checkout systems are also demanded by a considerable proportion of digital natives and can make shopping easier and more efficient. The use of other technologies, which are requested by a smaller proportion of consumers, can also be examined depending on the use case. The connectivity of distribution channels and the adoption of various technologies in the retail industry also enables the collection and analysis of customer-related data. This can contribute in various ways to individualizing and thus further improving the over-the-counter shopping experience.

5 Discussion

The previous chapter showed the results of the study and explained Generation Z's expectations of the stationary retail sector regarding digital services. The expectations of Generation Z differ from those of the other generations, which indicates a need for action for stationary retailers. In addition, technological advances have contributed to a shift in the consumer decision-making process and the touchpoints between retailers and consumers (Zöller, 2019). Due to the different customer journeys of target groups and generations, it is important to create, observe and link digital and physical sales and information channels along the three phases of the customer journey (Kleinjohann & Reinecke, 2020).

This does not mean compulsively and hastily digitizing everything, but rather looking at customers and their requirements on the stationary retail sector. As pointed out in the previous chapter, consumers do not expect pure online retail or stationary retail, but a mix between these two concepts, an omni-channel strategy (Pinker, 2018). Smart services can help to implement this strategy and to create digital touchpoints along the customer journey (Cui et al., 2022). Figure 2 presents relevant digital services and technologies for retailers along these phases to increase benefit and in turn attract Generation Z customers. In some cases, smart services cannot be assigned to just one phase of the customer journey and can therefore be found in more than one phase. Furthermore, the smart services are listed according to the importance for Generation Z determined from the survey. In addition, the technologies were highlighted by color according to the location of the touchpoints.

Starting with the first phase, the pre-purchase phase, Generation Z does not see digital/interactive shop windows as being quite as important. Nevertheless, this technology represents an interesting solution, as retailers can attract prospective customers to the store even after closing time. In the pre-purchase as well as in the purchase phase, Generation Z sees connectivity in the store with digital interactive terminals combined with extended product information on displays as a medium important requirement. Nevertheless, smart services based on digital signage can play a major role here. With this technology, retailers could also display goods digitally and would therefore not need to stock the entire product range. Likewise, the screens could act as digital price tags, allowing goods that are perishable to be sold more quickly. This would further strengthen sustainability efforts of retailers. One of the most important issues is that Generation Z sees the most important service

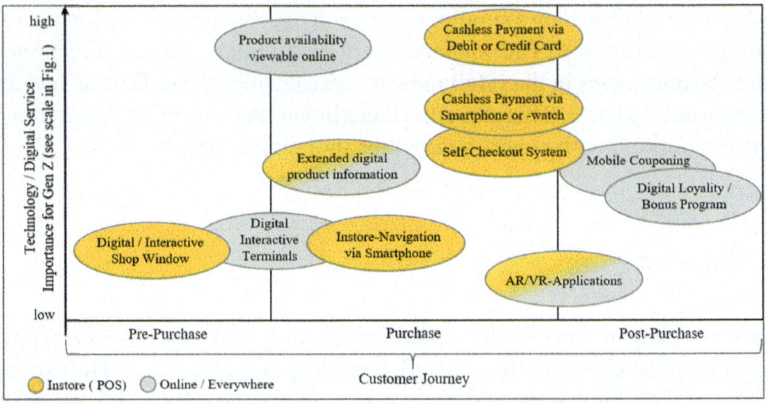

Fig. 2 Relevant Technologies for retailers along the customer journey to attract Generation Z customers

in the online availability of goods viewed in an online store and the link to social media. Consumers can save unnecessary trips to the retailer's location by checking the availability of goods online. This method pays off positively in terms of environmental sustainability as well (Deckert & Wohllebe, 2021). In the purchase phase and going further to the post-purchase phase, AR and VR applications can enhance the shopping experience of the customers. Nevertheless, the number of respondents who wanted an AR and VR solution as a touchpoint with their retailers was rather low. Regarding the state of the art, smart mirror technologies or 360° tours will however play a greater role in the upcoming years. Even more significant for Generation Z is the payment process. Cashless payment is indispensable for Generation Z in the retail sector and is a decisive factor in whether the customer comes back to the store or not. Generation Z can imagine using instore-navigation with their smartphone as well as paying at self-checkout systems and fully automated checkouts. This type of shopping is more stress-free for the customer, and, in addition, it helps the retailer to create value in the sense of creating beneficial services as referred to in the S-D logic. In the post-purchase phase of the customer journey, Generation Z wants to stay connected with the retailer through mobile couponing, a feedback system and digital loyalty/bonus programs. This would ensure that the customer journey gets repeated by satisfied customers. Further, smart and digital technologies make it possible to place touchpoints along the entire customer journey and wherever customers are. This can lead to an almost unmanageable number of touchpoints. But this also creates opportunities, as retailers can transfer their content to the right person at the right time and to get resilient (Zöller, 2019).

The collected data does not allow any reliable conclusions to be drawn from the sample to the population. A reason for this is that no quota characteristics of the sample were defined and considered. Therefore, the study does not claim to be representative. For future research, the consideration of further factors such as gender, educational level and income of the survey participants, as well as a further subdivision of the retail sector, could lead to interesting findings and more precise, target group-oriented recommendations for action. This paper brings a new approach and updated data to the academic discourse on how the important industry of retail should change to remain attractive to customers. Future research could therefore also look at the extent to which the retail sector is already addressing the demands of young consumers and meeting them with new innovative ideas and services. A further approach could result from a longitudinal view by comparing the requirements of each generation in pairs and anticipating future expectations.

6 Conclusion

The study shows that, in the context of digitization, there is a need for action on the part of traditional stationary retailers to satisfy the consumer needs of Generation Z. As Pinker states "retail dynamics are changing and the key task for retailers is to respond" (Pinker, 2018). Therefore, the optimal design of the customer journey from the initial contact to the act of purchase and the follow-up will develop into the decisive factor for the survival of retail in the coming years. This journey must also be taken digitally and include an omnichannel strategy through smart services to attract Generation Z. The sales channels will merge, and retailers will be able to offer their customers a seamless customer experience. The channel change is not only interesting regarding the purchase, but also for the entire customer journey (Grewal & Roggeveen, 2020). Considering these aspects, companies from the stationary retail sector can be enabled to fulfill the shift from a product-driven point of view to a service- and experienced based approach as mentioned in the S-D logic. In conclusion, it should be noted that the results of the study can only be seen as a snapshot and that the requirements of young consumers are changing, to which the retail industry needs to adapt constantly.

References

Batsakis, G., Konara, P., & Theoharakis, V. (2022). Digital sales channels and the relationship between product and international diversification: Evidence from going digital retail MNEs. *Global Strategy Journal*. https://doi.org/10.1002/gsj.1465

Cui, X., Xie, Q., Zhu, J., Shareef, M. A., Goraya, M. A. S., & Akram, M. S. (2022). Understanding the omnichannel customer journey. *Journal of Retailing and Consumer Services, 65*, 102869. https://doi.org/10.1016/j.jretconser.2021.102869

Deckert, R., & Wohllebe, A. (2021). *Digitalisierung und Einzelhandel: Taktiken und Technologien, Praxisbeispiele und Herausforderungen* (1st ed.). Springer Fachmedien Wiesbaden GmbH; Springer.

Handelsverband Deutschland e. V. (HDE): *Zahlenspiegel 2022*. [Online]. https://einzelhandel.de/index.php?option=com_attachments&task=download&id=10681

Esch, F.-R., Neudecker, N., & von Einem, E. (2010). Mehrwert für die Marke durch Kommunikation schaffen. *Marketing Review St. Gallen, 27*(3), 8–14. https://doi.org/10.1007/s11621-010-0054-1

Grewal, D., & Roggeveen, A. L. (2020). Understanding retail experiences and customer journey management. *Journal of Retailing, 96*(1), 3–8. https://doi.org/10.1016/j.jretai.2020.02.002

Heinemann, G. (2021). *Intelligent Retail: Die Zukunft des stationären Einzelhandels*. Springer Gabler.

Klaffke, M. (2021). *Erfolgsfaktor generationen-management—roadmap für das personalmanagement in generationen-management*. In M. Klaffke, (Ed.), (pp. 3–45). Springer Fachmedien Wiesbaden.

Kleinjohann, M., & Reinecke, V. (2020). *Marketingkommunikation mit der Generation Z*. Springer Fachmedien Wiesbaden.

Maas, R. (2022). *Neueste Generationenforschung in ökonomischer Perspektive: Reichen Generation X, Y und Z zur Beschreibung der Wirklichkeit aus?* (1st ed.). Verlag W. Kohlhammer.

Merkle, W. (2020). *Erfolgreich im stationären Einzelhandel: Wege zur konsequenten Profilierung im digitalen Zeitalter* (1st ed.). Springer Gabler.

Naskrent, J., Stumpf, M. & Westphal, J., Eds. 2021. *Marketing & Innovation 2021: Digitalität—die Vernetzung von digital und analog*. Springer Gabler. [Online]. Available https://swbplus.bsz-bw.de/bsz1688879102cov.htm

Pinker, A. (2018). *Innovationen im Handel. Innovationstechnologien und digitale Strategien für Einzelhändler*: disserta Verlag.

Sivapalan, A., Shankar, A., Das, M., & Konalingam, K. (2022). How to enhance customer loyalty to retail stores via retail service quality practices? *A Moderated Mediated Mechanism, Services Marketing Quarterly, 43*(4), 445–465. https://doi.org/10.1080/15332969.2021.1997504

Thaichon, P., Phau, I., & Weaven, S. (2022). Moving from multi-channel to Omni-channel retailing: Special issue introduction. *Journal of Retailing and Consumer Services, 65*, 102311. https://doi.org/10.1016/j.jretconser.2020.102311

Vargo, S. L., & Lusch, R. F. (2016). Institutions and axioms: An extension and update of service-dominant logic. *Journal of the Academy of Marketing Science, 44*(1), 5–23. https://doi.org/10.1007/s11747-015-0456-3

Wang, H. A., Yin, X., & Ouyang, A. Y. (2022). Store promotion, manufacturer competition and welfare. *Singapore Economic Review., 67*(04), 1479–1496. https://doi.org/10.1142/S0217590821500442

Zöller, S. (2019). *Die Customer Journey in Ja zur Digitalisierung!*. In: S. Zöller, (Ed.), (pp. 175–185). Springer Fachmedien Wiesbaden.

Remanufacturing and Product-Service Systems for the Circular Economy: A Business Model Analysis

Benedetta Franceschi, Gianmarco Bressanelli, and Nicola Saccani

Abstract Climate change and environmental pressures are pushing companies to adopt Circular Economy (CE) business models based on remanufacturing and on the provision of Product-Service Systems (PSS). However, specific guidelines on how to implement these business models are lacking. Thus, this paper aims to analyze CE business models based on remanufacturing and PSS, to define a full list of key recurrent and common elements that are needed to setup these specific business models. A multiple case study analysis is carried out, and a full list of configuration options for both remanufacturing and PSS business models is drafted. Results can help manufacturing companies in reshaping their value proposition, value delivery, value creation and value capture towards a CE based on remanufacturing and PSS.

Keywords Remanufacturing · Product-service systems · Business models · Circular economy

1 Introduction

Europe is pushing Circular Economy (CE) as a new economic solution to keep products, components, and materials at their highest value (Ellen McArthur Foundation, 2013). It decouples economic growth from resource extraction, since it aims to implement closed-loop production systems where products, components and materials are used repeatedly (Bressanelli et al., 2019). Several business models allow companies to achieve circularity. Among them, remanufacturing and product-service system (PSS) solutions (such as leasing or pay-per-use) are good candidates, since they both allow to reach environmental benefits and economic advantages, tuning components into high-quality and functionality products (Chierici & Copani, 2016). Despite this relevance, companies are struggling to change their linear approach and adopt these new CE business models. Looking at the literature, guidelines are missing. Thus,

B. Franceschi · G. Bressanelli (✉) · N. Saccani
RISE Laboratory, Department of Mechanical and Industrial Engineering, Università Degli Studi Di Brescia, Via Branze 38, 25123 Brescia, Italy
e-mail: gianmarco.bressanelli@unibs.it

J. Meierhofer et al. (eds.), *Smart Services Summit*, Progress in IS,
https://doi.org/10.1007/978-3-031-36698-7_14

this paper aims to analyze circular business models based on remanufacturing and PSS, to define a full list of key recurrent and common elements (defined hereafter 'configuration options') needed to setup specific business models. A multiple case study analysis combined with the Business model Canvas tool has been used to reach the research goal. Therefore, Sect. 2 provides the theoretical background on remanufacturing and PSS business models for the CE. Section 3 outlines the methodology, while results are presented in Sect. 4 and discussed in Sect. 5. Lastly, conclusions are presented in Sect. 6.

2 Theoretical Background

Remanufacturing in a CE aims to collect products that the consumer wants to dispose of, to upgrade them and extend their life (Khan et al., 2018; Sundin & Bras, 2005). By doing so, it avoids materials and energy consumption for manufacturing new components, reducing their environmental impacts and their costs. CE business models based on remanufacturing has been explored in several ways in literature. However, and despite their perceived benefits, they have not yet achieved a widespread application in industry (Jensen et al., 2019). Lüdeke-Freund et al. (2018) conducted a morphological analysis of literature to identify the major business model design options and define the potential strategies to support their development. Among these patterns, refurbishment and remanufacturing enable the comprehensive overhaul of products by replacing the failed parts, and they help companies in creating value using resources multiple cycles and reducing waste and consumption. To create this regenerative system, the authors suggest companies developing slowed (e.g., through long-lasting design for products), narrowed (e.g., through resource efficient use), and closed (e.g., through resource loop) resource flows. A higher level of analysis was conducted by (Nußholz, 2017). She tried to frame the circular business model research field, by defining what a circular business model is from the perspectives of resource efficiency and business model innovation. The analysis was conducted under the 'Business Model Canvas' theoretical lens, but remanufacturing appears only as an example of a strategy to extend product useful lifetime. Lastly, (Jensen et al., 2019) explored how an integrative perspective can drive sustainable value creation in the remanufacturing business model context, by the limited perspective of Original Equipment Manufacturers (OEMs). All the analyzed company cases demonstrate that remanufacturing brings different benefits in terms of resource efficiency, costs, and job opportunities. Moreover, remanufacturing business models are often combined with PSS as an enabling factor (Yang, 2018). Indeed, PSS is considered a promising CE strategy, since it can change the way of creating value: manufacturers or retailers own products and enable a 'functional service model', while costumers adopt a new 'consume and return' pattern, through the access to a service or a product use (Sundin & Bras, 2005). Thus, PSS definition includes concepts such as servitization, service-dominant logic, pay-per-use services and outcome-based services (Brodie et al., 2019). Sousa-Zomer et al. investigated a single case study to find out the challenges of implementing a circular business model from a renting perspective, emphasizing the need to combine tangible and intangible business aspects in the value creation, such as by changing the organizational mindset, creating collaborative circular networks with stakeholders and new partners, or offering a structured after-sales service (Sousa-Zomer et al., 2018). Through the exploration of a Chinese state-owned company for air separation units, Yang et al. (2018) stated that PSS business models can generate value through the circularity of supply chains. The analysis compared product-, use- and result-oriented PSSs, crossed with their way of creating value. Findings demonstrated that the regular provision of technical services of maintenance, repair and remanufacture can push towards a more circular supply chain. The study also shows that the closer the value creation is to result-oriented PSS, the tighter and more efficient the cycle is, since products need less intervention to come back to cycle again, and the higher potential environmental and economic savings are (Yang et al., 2018).

3 Research Methodology

To accomplish the paper objective, company case studies have been analyzed according to the Business Model Canvas tool (Osterwalder & Pigneur, 2010), chosen because it allows a simple and intuitive representation of how a company *creates*, *delivers*, and *captures* value, despite the complexity of elements that affect the company operations. The tool is composed of several interlinked blocks, grouped into the value proposition, the value delivery (customer segments, customer relationships, and channels), the value creation (key resources, activities, and partnerships), and the value capture (costs and revenues). Four companies (Table 1) have been chosen as representative case studies for the remanufacturing or pay-per-use business models. RMN1 and RMN2 case studies are based on primary sources (i.e., interviews), while PSS1 and PSS2 case studies are based on secondary information.

Table 1 Case studies

Case study	Industry	Turnover (€)	Employee	Region
RMN1	Sport fitness equipment	600 million	2000	Italy
RMN2	Household appliances	15 million	60	Italy
PSS1	Tire manufacturing	24,13 billion	127.187	France
PSS2	Textiles	822.000	7	Netherlands

4 Case Studies

RMN1 is the Italian leader in the fitness and wellness sector. It produces sport and fitness equipment and, recently, the company started offering a remanufacturing service on its own old products, keeping the quality standards by inspecting, testing and, if needed remanufacturing the original spare parts. Regarding *the value proposition*, RMN1 offers its customers the possibility to buy an updated and high-quality fitness equipment at a lower price. Concerning the *value delivery*, it is related to the opportunity for customers to buy a fitness equipment and practice gym at home. However, fitness equipment is usually expensive: customers need an appropriate storytelling to get involved and to understand that remanufactured products are cheaper. The *value is created* by designing gym equipment in a modular way, to enhance remanufacturing opportunities. Once products reached the end of life, reverse logistics is activated to collect machines and stocked them in a warehouse to be inspected and evaluated eligible for remanufacturing. In case of eligibility, the equipment is remanufactured directly in the headquarter, where the key activities, such as the disassembly of the useful components, the management of spare parts, the reassembly, and final quality tests, are internally carried out. Lastly, *value is captured* by selling remanufactured products, which maintain a fair value over the time because of long-term contractual agreements for upgrading, while costs are mainly related to reverse logistics and the remanufacturing process. RMN2 is an Italian distributor of household appliances spare parts. Thanks to its technical experience on components and spare parts, the company started in 2017 a project of remanufacturing old appliances, in a way to also create new job opportunities. Concerning the *value creation*, RMN2 makes a broken appliance work again at 50% the price of a new machine. Moreover, customers positively contribute to both the society and the environment since they reduce waste and increase job opportunities. *Value is delivered* by addressing to three categories of customers: people who cannot afford new expensive appliances, those who live in temporary or second house, or those interested in tackling environmental issues. They are all reached through physical and digital channels, including the company website. While concerning the relationship between the company and customers, it is usually limited to the purchase, including 12 months of warranty and technical support. *Value is created* through the collection of old appliances by RMN2 in collaboration with a logistics operator, but anyone can also donate his end-of-life appliance via website. Then, collected devices are brought to a laboratory to be checked and broken components are replaced with OEM spare parts, leveraging on RMN2 long-time experience and technical knowledge. Then, appliances are tested and sanitized. Lastly, *value is captured* by sales of remanufactured products, which covers the collection and remanufacturing process costs, especially related the workforce. PSS1 is a multinational company which manufactures and sells tires. To improve its environmental footprint, it introduced a product as a service pilot project, by providing tires and all the associated services such as tires management and maintenance, while customers pay a fee based on kilometers driven in a time-fixed period. Moreover, offering an additional service of usage monitoring (tires temperature and pressure) thanks to Internet of Things (IoT) sensors, PSS1 can suggest a proper driving, reducing tires wear, replacement and costs, and increasing environmental benefits. So, PSS1 *creates value* by providing customer the tires and services for a monthly fee based on the distance covered. The company also offers additional services such as drivers' training and personalized advice, to guarantee tires durability, fuel efficient usage and environmental benefits. Concerning *value delivery*, the project is targeted to customers who covers long distance with their van and truck fleets. To reach them, the company create a long-term relationship by offering support and the previous personalized services. *Value is created* through a lifecycle approach: PSS1 handles every aspect from tire selection to maintenance and support, retreading and recycling, thanks to IoT sensors and data analysis. These monitoring systems help to optimize preventive maintenance, avoid incidents and downtime. Lastly, *value is captured* through the pay-per-use fee and the additional services sales, which covers the costs of tires manufacturing, management, and maintenance. PSS2 is a fashion startup company that leases sustainable jeans—made with certified organic cotton and recycled fabric—to reduce waste of 'fast fashion' industry, especially generated during their production. Customers can rent jeans for one year, at the end of which they can decide to keep the jeans, give them back, or swap them with other models. Thus, *value proposition* is focused on one-year jeans leasing. At the end of the contract, the customer can decide to swap with new jeans to always wear up-to-date clothing, but with less environmental impact. The jeans level of sustainability also exceeds the competition, because of their organic cotton and recycled fabrics and their long-lasting design. Indeed, free repairs are included for all the contract duration. *Value is delivered* by involving people sensitive to the environmental impact of the fashion industry and to sustainability. So, PSS2 invests a lot on web communication and storytelling. Finally, *value is created* from the jeans collection at the end of contracts, since it designs the products to last and to be suitable for all the seasons, so that customers wear jeans for as long as possible. Their production process is carried out by a network of partners. First, recyclers collect old jeans. Then, yarns are dyed and weaved into fabrics, which are cut and stitched to create the final product. The logistics is managed externally, while selling key activities (marketing, sales, and customer service) are internal. Lastly, *value is captured* from the leasing monthly fees, which cover the costs of jeans design, production, logistics and service activities.

5 Discussion

In this section, a comparison and a list of the relevant features of both remanufacturing and PSS circular business models are provided (Fig. 1).

Value proposition in remanufacturing business model is encapsulating in the offering of 'as good as new' products, allowing customers to buy high-quality products as much as the new ones at an affordable price, as shown by RMN1 and RMN2. In addition, these products are often offered with a warranty period at least long as new

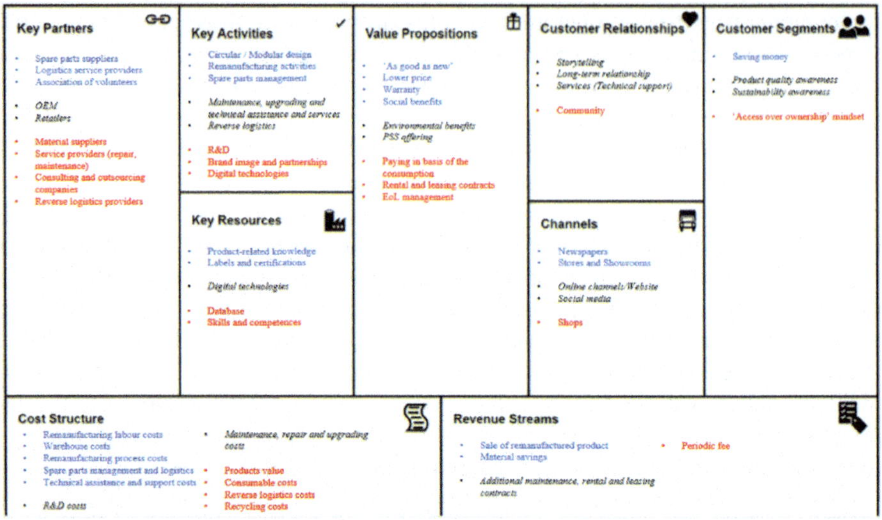

Fig. 1 Business Model Canvas highlighting remanufacturing (blue), PSS (red) and common elements (black) configuration options

ones, which is a way to attract customers. However, from the other side it is a risk for the company, which must cover all costs in case of failure. Differently, PSS models *propose value* by focusing on the provision of product 'as a service', following a leasing or pay-per-use scheme. This leads to an added value for customers, since it reduces the initial investment related to a product purchase (PSS1). In addition, a series of services are guaranteed for all the contract duration, such as maintenance, repair, and end-of-life management. Lastly, both the business models introduced contribute to environmental and social benefits. Remanufacturing business models aim to extend products lifespan and reduce waste and raw materials usage. Meanwhile, they provide social impact by creating job opportunities. In PSS business models, however, sustainability can be improved in different ways, depending on the business case. In PSS1, some companies provide long lasting, high quality and high-efficient products, and others focus on consumables reduction during usage. While others concentrate on the collection and recycling of leased products at the end of life (PSS2).

Value is delivered to different customer segments, depending on business models. Remanufacturing business models target customers who care about product quality and are looking for saving opportunities. So, to spread awareness about 'as good as new' remanufactured products among customers, companies must invest in communication plans and brand awareness, to show their economic benefit, especially for usually expensive products (RMN2). PSS business models mainly address costumers that have an 'access over ownership' mindset and that occasionally use the product, by optimizing the user's cost compared to a high initial investment and a low rate of product utilization (PSS2). Moreover, it optimizes costs of companies asking for products that requires periodical maintenance (PSS1). Unlike the traditional ones, remanufacturing and PSS business models try to create a long-term relationship with costumers thanks to specific agreements and/or services offered (such as warranty, maintenance, etc.). Communication, and in particular storytelling, is a useful tool for that purpose, as it allows to highlight the environmental and social benefits. In this way, both CE business models mainly use online channels (e.g., dedicated website pages, blog, or social medias) where they publish news or articles about them or about environment (PSS2). Physical channels (e.g., stores, showrooms) are also important to allow customers to see the product quality personally. For instance, RMN2 has shops and laboratories in Italy where products and remanufacturing process are shown.

Value is created thanks to several resources. For remanufacturing business models, technical staff knowledge is essential to know how to manage products for disassembly, diagnosis, reassembly, etc. For instance, RMN2 knows spare parts management in depth. Digital technologies are equally relevant, mainly for providers of PSS and additional services (maintenance and repair). In PSS models based on pay-per-use, customers' usage is monitored to calculate the right amount of the fee (PSS1). Moreover, by collecting and analyzing data, companies can advise an optimistic usage of their products. Lastly, environmental label is another key resource, to prove the sustainability and quality of the products-services offered. In remanufacturing business models, circular modular design is a key activity to enable products disassembly and reassembly (RMN1). Another important key activity is reverse logistics. Companies have to manage the collection of broken products to be remanufactured or leased ones at the end of subscriptions. Thus, a key partner is a third-part reverse logistics service provider. Spare parts supply is critical too in remanufacturing products to be 'as good as new'. So, company often store many components on their own (e.g., RMN2). Key partners are essential for the success of both remanufacturing and PSS business models. In general, remanufacturing companies do not coincide with OEM manufacturer, but partnerships are established to guarantee the quality of remanufactured products. Lastly, both remanufacturing and PSS business models often have 'green' partnerships with associations, foundations, or non-profit organizations to guarantee the achievement of environmental and social objectives (RMN2, PSS2).

Lastly, *value is captured* from the sales of remanufacturing products (in remanufacturing business models), which have a lower price due to reused components, and, in some cases, from additional services (maintenance, leasing contracts). In PSS models, revenues mainly come from the periodic fee paid by costumers, usually coupled with an initial deposit (PSS1 calculates the fee based on kilometers driven, PSS2 jeans are paid on a fixed fee). The most relevant cost of remanufacturing business models is represented by the labor cost. Indeed, most of repairing and overhauling activities can be carried out only manually. Disassembling, checking the components, replacing parts, and reassembling products are rarely automated, and need skilled technicians (in RMN2 almost 10 employees are focused on remanufacturing). On the contrary, PSS business models have to bear the initial product cost, which represents an anticipate and risky investment. Another important cost is reverse logistics, which does not include only the transportation cost, but also inventory, spare parts, etc.

6 Conclusion

This research fills the lack of guidelines in literature on common configuration options for circular business models such as remanufacturing and PSS. By outlining a full list of configuration options, companies can be facilitated in their application. Each case is described and analyzed using the Business Model Canvas tool, both for remanufacturing and pay-per-use business models. This tool enabled the comparison among the cases and the business models to find the elements which are the same time more common and representative. However, both the business models highlight the way to reach a more circular path. The models show how costumers can be drawn by a high-efficient and high-quality second-hand product, either in buying it at a lower price (remanufacturing) or in obtaining it as a service (pay-per-use, leasing, rental). Both models also show how companies should change their business configuration, addressing towards activities such as modular design, in-depth technical expertise, in return of an environmental and, hopefully, economic benefit. Since case studies come from different industries, we can potentially argue that our findings have a high degree of generalization. However, more case studies should be carried out, to further improved and validate our business model framework.

References

Bressanelli, G., Perona, M., & Saccani, N. (2019). Challenges in supply chain redesign for the Circular Economy: A literature review and a multiple case study. *International Journal of Production Research, 57*, 7395–7422. https://doi.org/10.1080/00207543.2018.1542176

Brodie, R., Löbler, H., & Fehrer, J. (2019). Evolution of service-dominant logic: Towards a paradigm and metatheory of the market and value cocreation? *Industrial Marketing Management, 79*, 3–12. https://doi.org/10.1016/j.indmarman.2019.03.003

Chierici, E., & Copani, G. (2016). Remanufacturing with Upgrade PSS for New Sustainable Business Models. *CIRP Journal of Manufacturing Science and Technology, 29*, 245–256. https://doi.org/10.1016/j.procir.2016.03.055

Ellen McArthur Foundation. (2013). *Towards a circular economy: Business rationale for an accellerated transition*

Jensen, J., Prendeville, S., Bocken, N., & Peck, D. (2019). Creating sustainable value through remanufacturing: Three industry cases. *Journal of Cleaner Production, 218*, 304–314. https://doi.org/10.1016/j.jclepro.2019.01.301

Khan, M., Mittal, S., West, S., & Wuest, T. (2018). Review on upgradability—A product lifetime extension strategy in the context of product service systems. *Journal of Cleaner Production, 204*, 1154–1168. https://doi.org/10.1016/j.jclepro.2018.08.329

Lüdeke-Freund, F., Gold, S., & Bocken, N. (2018). A review and typology of circular economy business model patterns. *Journal of Industrial Ecology, 23*, 36–61. https://doi.org/10.1111/jiec.12763

Nußholz, J. (2017). Circular business models: defining a concept and framing an emerging research field. *Sustainability, 9*, 1810. https://doi.org/10.3390/su9101810

Osterwalder, A., & Pigneur, Y. (2010). *Business model generation: a handbook for visionaries, game changers, and challengers*

Sousa-Zomer, T., Magalhães, L., Zancul, E., & Cauchick-Miguel, P. (2018). Exploring the challenges for circular business implementation in manufacturing companies: An empirical investigation of a pay-per-use service provider. *Resources, Conservation and Recycling, 135*, 3–13. https://doi.org/10.1016/j.resconrec.2017.10.033

Sundin, E., & Bras, B. (2005). Making functional sales environmentally and economically beneficial through product remanufacturing. *Journal of Cleaner Production, 13*, 913–925. https://doi.org/10.1016/j.jclepro.2004.04.006

Yang, M., & S. P. (2018). Product-service systems business models for circular supply chains. *Production Planning & Control, 29*, 498–508. https://doi.org/10.1080/09537287.2018.1449247

Yang, M., Smart, P., Kumar, M., Jolly, M., & Evans, S. (2018). Product-service systems business models for circular supply chains. *Production Planning & Control, 29*, 498–508. https://doi.org/10.1080/09537287.2018.1449247

Sustainable Smart Services for Financial Institutions: A Framework to Support Banks in ESG Integration in the Risk Analysis Process of Loans

Sabrina Urban, Nicola Moosbrugger, Michael Hellwig, and Martin Dobler

Abstract Through mandatory ESG (environmental, social, governance) reporting large companies must disclose their ESG activities showing how sustainability risks are incorporated in their decision-making and production processes. This disclosure obligation, however, does not apply to small and medium-sized enterprises (SME), creating a gap in the ESG dataset. Banks are therefore required to collect sustainability data of their SME customers independently to ensure complete ESG integration in the risk analysis process for loans. In this paper, we examine ESG risk analysis through a smart science approach laying the focus on possible value outcomes of sustainable smart services for banks as well as for their (SME) customers. The paper describes ESG factors, how services can be derived from them, targeted metrics of ESG and an ESG Service Creation Framework (business ecosystem building, process model, and value creation). The description of an exemplary use case highlighting the necessary ecosystem for service creation as well as the created value concludes the paper.

Keywords ESG · Sustainable smart service · S-D logic · SME · Financial institutions

---

S. Urban (✉) · M. Hellwig
Vorarlberg University of Applied Sciences, Josef Ressel Centre for Robust Decision Making, Hochschulstraße 1, Dornbirn, Austria
e-mail: sabrina.urban@fhv.at

M. Hellwig
e-mail: michael.hellwig@fhv.at

N. Moosbrugger · M. Dobler
Vorarlberg University of Applied Sciences, Research Centre Business Informatics, Hochschulstraße 1, Dornbirn, Austria
e-mail: nicola.moosbrugger@fhv.at

M. Dobler
e-mail: martin.dobler@fhv.at

J. Meierhofer et al. (eds.), *Smart Services Summit*, Progress in IS,
https://doi.org/10.1007/978-3-031-36698-7_15

1 Introduction

New EU regulations force banks to take ESG factors into account when granting loans and assessing portfolios by 2026. Since a lot of the necessary data is not available comprehensively, or exists at all, banks must deal with this topic more intensively. The identification and extraction of relevant data from different sources to transform, combine, and evaluate the data into insights and actions that will be considered for credit risk assessment for companies is essential. For financial institutions, the value added of including ESG policies into their strategies may not be obvious, but the long-term benefits can be equally significant (La Torre et al., 2021). By prioritizing loans to, and investments in companies that consider ESG factors in the screening process, banks can finance more robust projects and companies with more stable profitability (Ioannou & Serafeim, 2019). This whole issue has been investigated through a smart science approach that focuses on potential value effects of sustainable smart services for both banks and their SME clients. This research builds on already conducted investigations on smart services for circular economy in the financial sector (Suzic et al., 2022).

2 State of the Art

To establish a link between service science ideas and the ESG subject, this chapter summarizes current findings from these two areas.

2.1 ESG

The abbreviation ESG refers to the environmental, social, and governance factors that affect the measurement of the responsibility and sustainability of a company's actions. Environment considers for example developments in climate protection, investment in renewable energy, and resource scarcity (EBA, 2022). The field social covers areas such as employee protection, health and safety, or equal opportunities (EBA, 2022). Governance, in contrast, deals with corporate ethics, risk management, and compliance (EBA, 2022). In recent years, the consideration of corporate ESG risks on the credit risk became increasingly important for the financial sector, as mandatory reporting grows on the part of the European Central Bank (ECB). The integration of ESG data is mandatory for banks, the rating agencies focus on the rating of companies. However, they focus on data from companies that have a disclosure obligation due to their size or legal status. The inclusion of data from SMEs is challenging for financial institutions, as current disclosure obligations do not apply to SMEs. ESG data of SMEs is relevant for financial institutions to establish a consistent basis for the assessment of their customer portfolios. This requires banks to collect sustainability data of their SME customers independently to ensure complete ESG integration in the risk analysis process for loans. ESG information is of great importance for both banks and SMEs. Financial institutions can establish an additional business model and sell the existing data to their customers. With this information, SMEs can learn their ESG risk assessment in advance, take action and make adjustments to improve their credit rating. Hence, through collaboration, both sectors can benefit from each other.

2.2 *S-D Logic, Service Science for Sustainability*

In service science and especially in service dominant logic (S-D logic), environmental sustainability has not been considered extensively so far (Vargo & Lusch, 2017). S-D logic, originally a marketing theory, was introduced by Vargo and Lusch (2004). Through elven foundational premises (FP) and five axioms (AX) the concept explains the shift form a goods dominant to a service dominant perspective, considering service as the basis of exchange (FP1) (Vargo & Lusch, 2004, 2008, 2016). Over the years scholars from all over the world and different fields of research have adapted and refined the theory, creating a more holistic approach (Vargo & Lusch, 2016). Wolfson et al. (2010) argue that the broadening of scope in service science shifted the focus from economically measurable services to more operational and responsible values. Thus, topics such as social and environmental sustainability gained more attention and are put at the core of corporation's value creation activities (Hansen & Grosse-Dunker, 2013). According to Vargo and Lusch (2016) value is co-created, always including the beneficiary (AX2). While co-creating a sustainable service, the value-in-use perspective is beneficial, since it enables the creation of value in multiple dimensions (Wolfson et al., 2010).

3 Problem Description

The European Green Deal has the goal to make Europe the first climate-neutral continent by 2050. In this regard, this includes the financial sector as key driver, as set out in the Commission's Action Plan on Financing Sustainable Growth (European Central Bank, 2020). The process of shifting to a low-carbon and more sustainable economy creates both risks and opportunities for the economy and financial institutions (European Central Bank, 2020). The ECB formulated a guide on climate and environmental risks that outlines the expectations regarding managing such risks and their respective disclosure requirements (European Central Bank, 2020). The main drivers of climate-related and environmental risks are classified as physical or transition risks. As Fig. 1 shows, environmental risks are subdivided into physical and transition risks. Acute physical risks include climate and weather-related events that occur unexpected. Chronic events are ongoing changes in climate patterns (EBA,

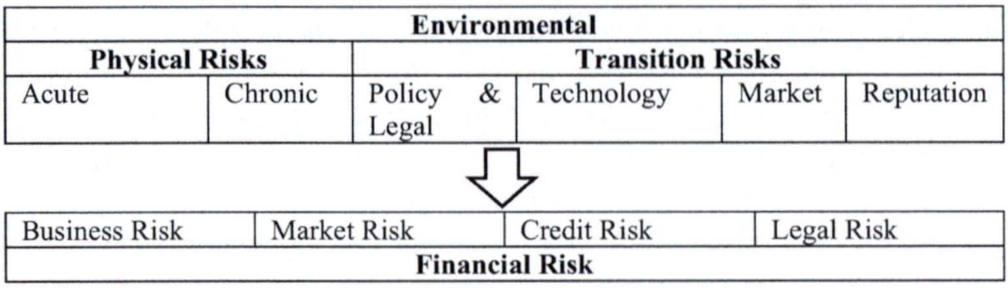

Fig. 1 Classification of climate related risks (EBA, 2022)

2022). Sectors that are most likely to be impacted by physical risks include agriculture, energy, transport, or tourism. Whereas transition risks describe the impact of technologies, policy and regulations or reputation of a company. The industries most affected by the transition to a zero-carbon economy contain energy, transportation, and manufacturing, whereas any climate threat has an impact on the existing financial risks of banks.

Currently, the focus is on the integration of environmental risks, but social as well governance risks are not negligible and potential metrics focus on environmental, social, and governance factors. Environmental factors include emissions, energy efficiency, water consumption, and waste production. Social factors describe which standards apply to employees, relationships with customers, and human rights in the company. The governance factors focus on a company's strategy and risk management, ethical considerations, as well as transparency. To measure the risks indicators, metrics are defined that are used in the different areas. The metrics described in Table1, provided as examples by the Task Force on Climate-Related Financial Disclosures (TCFD), are considered in the environmental risks category whereas metrics covering transition risks contain the amount and scope of assets or business activities exposed to transition risks (EBA, 2022; Task Force on, Climate-related and Financial Disclosure, 2021).

Currently, ESG data is not uniformly available, which causes a barrier for financial institutions to integrate the metrics into the ESG scoring approach or the credit risk

Table 1 Exemplary Metrics for ESG risk indicators

Metric domain	Potential measure unit	Example metrics
Green House Gas (GHG) Emission	Tons CO_2 equivalent	Absolute Scope 1, Scope 2, Scope 3 GHG emissions
		Weighted average carbon intensity
		GHG emissions per MWh of electricity produced
Transition risks	Percentage	Volume of real estate collaterals highly exposed to transition risk
Physical risks	Percentage	Proportion of property, infrastructure, or alternative asset portfolios in an area subject to flooding, heat stress, or water stress

assessment, respectively. Clear and comparable ESG datasets are necessary but both individual data from large enterprises and (either individual or collective) data from SMEs is rarely publicly available. Creating a dedicated foundation to collect data is therefore essential as most ESG data is not existing for regional, non-listed companies. This is caused by the fact that regional, local companies usually do not have disclosure obligations, which can lead to major challenges for financial institutions to generate, analyse, and assess ESG data for their clients. Various studies have noted significant deficiencies with the methods and the ongoing disorganisation of the data (Bose & Springsteel, 2017; Cort & Esty, 2020). Many of the instruments, frameworks, and mechanisms for measuring corporate sustainability are lacking sufficient depth to be applicable in the banking sector (Bose & Springsteel, 2017). Additionally, it has been pointed out that poor organisation of data and insufficient tools (Cort & Esty, 2020) are challenges which need to be addressed. Companies' ESG ratings are typically done using data from ESG rating agencies – public institutions that evaluate companies' sustainability performances using their own methods. To address these challenges, data science approaches can be used and incorporate metrics for SMEs in lending. Statistical methods can be applied to find correlations within the data to identify and complete the data set with relevant characteristics. Further down the line, it is possible to use machine learning techniques to find relationships between features and thus integrate them into the scoring system.

4 Developing an ESG Service Creation Framework

In order to generate a service that creates value for the actors involved, an innovative and creative service creation process needs to be gone through. To shape this process in an efficient and effective way structural guidance is advantageous and often necessary. Therefore, an *ESG Service Creation Framework* to visualise and structure the process is suggested. The framework is deduced from the Business Ecosystem Canvas (Burkhalter, 2020) which is an adapted version of the Business Model Canvas (Osterwalder & Pigneur, 2011). The ecosystem approach is considered suitable since the research field of business ecosystems puts the centre of attention on revealing the dynamics and patterns of an ecosystem. Special focus is thereby on emphasising different types of business players and analysing the mechanism of their network – see Fig. 2 (Tsujimoto et al., 2018). Thus, the business ecosystem approach sets the right scene to identify the, according to S-D logic, multiple actors who are involved in the co-creation process of value (Vargo & Lusch, 2004, 2008, 2016).

The creative service creation process, as illustrated in Fig. 3 bases on a typical flowchart. Flowcharts visualize single steps that compose a process in a chronological order. Through their graphical nature, they provide a holistic overview and support easy understanding of processes (Chapin, 2003). That is why a simplified flowchart explaining the necessary steps for service creation has been outlined. Thereby, the

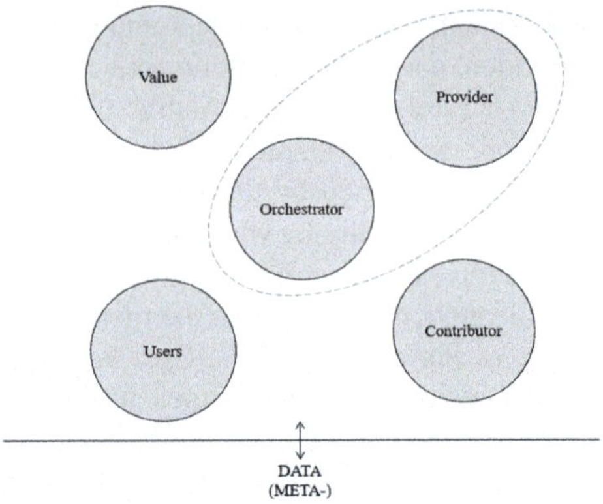

Fig. 2 Business ecosystem

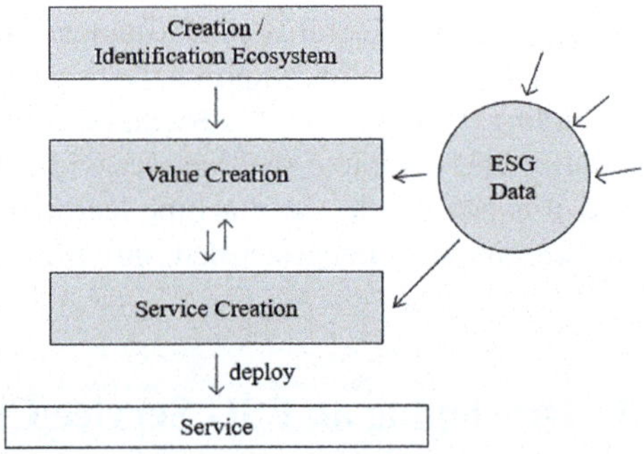

Fig. 3 Process model

coherences of the process steps as well as their iteration are highlighted. Additionally, input data gets specified and indicated where this data enters the process.

Through a combination of these two approaches the *ESG Service Creation Framework* was developed. The ESG Service Creation Framework consist of five main parts (main body), namely users, providers, contributors, orchestrators, and value. Thereby, the first four areas describe different types of actors that need to be identified in order to cover the multiple actors involved in the co-creation process of value. While depicting the actors, a respective use case (right side/bottom) builds the defining basis. The field value will most probably be defined during the process, can however also be defined in advance if a concrete idee or desired outcome already exists. The whole framework is built in a modular way to enable a flexible and individual scope for design, resulting in the possibility to highlight interconnections between the components of the fields. Therefore, it becomes easier to visualise an overview of the ecosystem, to emphasise gaps (e.g., missing/still needed actors, missing data in use case), and to derive possible value outcomes. Furthermore, the framework has been designed circularly, to encourage and support iterations. It is of crucial importance to go back and forth within the model or work through the whole process again after finishing it for the first time. The iteration helps with identifying gaps or blind spots, but also encourages new ideas, through which improved or additional services can be developed. As stated in S-D logic, value is co-created always including the beneficiary (AX2) and actors cannot deliver value, but can participate in the creation and offering of value propositions (FP7) (Vargo & Lusch, 2004, 2008, 2016). Therefore, it is recommended to include several different actors, who are, or might be involved in the later value creation or service process, during the development of the ESG Service Creation Framework.

5 Exemplary Use Case

The following business use case presents an example how the ESG Service Creation Framework (already pre-filled out in Fig. 4) can be applied in practice. After defining the use case, which gives the context for the following deliberations, the ecosystem which is affected by the use case needs to be defined. Contributors: With the mandatory inclusion of ESG ratings in the granting of loans, banks must collect relevant data from their customers. The choice of data which is collected depends on the guidelines specified by the EU. Since no mandatory disclosure exists for SMEs, voluntarily provided data of the SMEs must be enriched by data from NGOs, commercial data providers or publicly available data from a particular country. Finally, financial institutions have a variety of ESG related data about their customers (value). Users: This specific knowledge allows banks to evaluate their clients on their ESG activities (value). This holistic view of the use case gives already a clear view of potential services. Service: SMEs that significantly reduce their emissions and make future investments in this area receive a certificate that rewards their efforts. By selling ESG ratings from banks (provider) to their clients, SMEs have the possibility to compare their rating with other companies in the same industry sector and region (value). Consequently, the resulting service has been applied to the use case, which most probably creates new gaps and questions that need to be answered, hence the ecosystem must be reconsidered to identify every necessary actor.

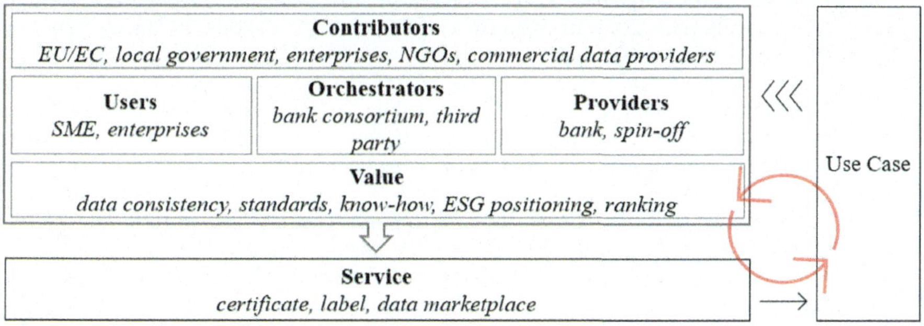

Fig. 4 ESG service creation framework

6 Outlook

The next stage will be to apply the ESG Service Creation Framework in a real-world setting. One method to do so would be through workshops that include all stakeholders of the ESG Service Creation Framework. Through this, individual participants can exchange ideas and discuss the process of the model as well as the practical application in detail. This enables the identification of required modifications and thereby the model can be improved within the individual steps. By testing the Framework additional use cases are likely to arise, which will broaden the view of its possibilities but also its limits. Therefore, a concrete description of the Framework and its usage is necessary in order to guarantee added value through the implementation of the Framework during the service creation process. Besides, more intense research regarding ESG as well as service science in banking industry and ESG integration is needed to gain more detailed and deeper knowledge, and to enhance the interaction of these topics.

Acknowledgements The financial support by the Austrian Federal Ministry of Labour and Economy, the National Foundation for Research, Technology and Development and the Christian Doppler Research Association is gratefully acknowledged.

References

Bose, S., & Springsteel, A. (2017). The value and current limitations of ESG data for the security selector. *Journal of Environmental Investing, 8*(1), 54–73.

Burkhalter, M. (2020). *Allocentric Business models—An d [Dissertation]*. University of St.

Chapin, N. (2003). Flowchart. In *Encyclopedia of Computer Science* (pp. 714–716). John Wiley and Sons Ltd.

Cort, T., & Esty, D. (2020). ESG Standards: Looming Challenges and Pathways Forward. *Organization & Environment, 33*(4), 491–510. https://doi.org/10.1177/1086026620945342

EBA. (2022). *EBA Report on management and supvervision of ESG risks for credit institutions and investment firms.* https://www.eba.europa.eu/sites/default/documents/files/document_library/Publications/Reports/2021/1015656/EBA%20Report%20on%20ESG%20risks%20management%20and%20supervision.pdf

European Central Bank. (2020). *Guide on climate-related and environmental risks.* 54.

Hansen, E. G., & Grosse-Dunker, F. (2013). Sustainability-Related Innovation. In S. O. Idowu, N. Capaldi, L. Zu, & A. D. Gupta (Eds.), *Encyclopedia of Corporate Social Responsibility* (pp. 2417–2417). Springer Berlin Heidelberg. https://doi.org/10.1007/978-3-642-28036-8_101771

Ioannou, I., & Serafeim, G. (2019). Corporate sustainability: A strategy? *Harvard Business School Accounting & Management Unit Working Paper, 19–065.*

La Torre, M., Leo, S., & Panetta, I. C. (2021). Banks and ENVIRONMENTAL, SOCIAL AND GOVERNANCE drivers: Follow the market or the authorities? *Corporate Social Responsibility and Environmental Management, 28*(6), 1620–1634. https://doi.org/10.1002/csr.2132

Osterwalder, A., & Pigneur, Y. (Eds.). (2011). *Business model generation: Ein Handbuch für Visionäre, Spielveränderer und Herausforderer*. Campus

Suzic, B., Urban, S., Hellwig, M., & Dobler, M. (2022). Smart circular economy value drivers: The role of the financial sector in stimulating smart regional innovation-driven growth. In S. West, J. Meierhofer, & U. Mangla (Eds.), *Smart services summit* (pp. 55–64). Springer International Publishing. https://doi.org/10.1007/978-3-030-97042-0_6

Task Force on, Climate-related, & Financial Disclosures (TCFD). (2021). *Guidance on Metrics, Targets, and Transition Plans*. https://assets.bbhub.io/company/sites/60/2021/07/2021-Metrics_Targets_Guidance-1.pdf

Tsujimoto, M., Kajikawa, Y., Tomita, J., & Matsumoto, Y. (2018). A review of the ecosystem concept—Towards coherent ecosystem design. *Technological Forecasting and Social Change, 136*, 49–58. https://doi.org/10.1016/j.techfore.2017.06.032

Vargo, S. L., & Lusch, R. F. (2004). Evolving to a new dominant logic for marketing. *Journal of Marketing, 68*(1), 1–17. https://doi.org/10.1509/jmkg.68.1.1.24036

Vargo, S. L., & Lusch, R. F. (2008). Service-dominant logic: Continuing the evolution. *Journal of the Academy of Marketing Science, 36*(1), 1–10. https://doi.org/10.1007/s11747-007-0069-6

Vargo, S. L., & Lusch, R. F. (2016). Institutions and axioms: An extension and update of service-dominant logic. *Journal of the Academy of Marketing Science, 44*(1), 5–23. https://doi.org/10.1007/s11747-015-0456-3

Vargo, S. L., & Lusch, R. F. (2017). Service-dominant logic 2025. *International Journal of Research in Marketing, 34*(1), 46–67. https://doi.org/10.1016/j.ijresmar.2016.11.001

Wolfson, A., Tavor, D., Mark, S., Schermann, M., & Krcmar, H. (2010). S^3—Sustainability and services science: novel perspective and challenge. *Service Science, 2*(4), 216–224. https://doi.org/10.1287/serv.2.4.216

Typology for Industrial Customers in the Subscription Economy

Günther Schuh, Christian Holper, Lennard Holst, and Volker Stich

Abstract More and more companies in the manufacturing industry are striving to transform from a transactional to a service-driven so-called "Subscription" business model to benefit monetarily from the opportunities offered by Industry 4.0. One hurdle in this transformation is the formulation of the appropriate service offering. While customers' demands for value provided with the service offering are becoming ever more extensive, the complexity of a combined service offering consisting of product, service, and software (product-service system) is also increasing. In practice as well as in the literature, recurring patterns of specific user types can be identified, which in turn help to standardize the service offering. However, this concrete user and value consideration has not yet been applied to business models in the subscription economy. The two central research methods that are used here are a detailed literature analysis and interviews with experts from the industry. The aim of this work is the identification of concrete customer types and their characterization based on specific features and attributes. With the help of specific customer types, companies in the manufacturing industry should be able to align their range of services in a targeted and user-oriented manner. In addition, the concrete addressing of identified customer types helps to better meet customer requirements and consequently to increase business success.

Keywords Subscription economy · Customer typology · Product-service system

1 Introduction

As industrial companies in the manufacturing industry are faced with increasing challenges due to globalization, increased competition, and a reduction of their innovation cycles, they are searching for new opportunities to improve their revenue and strengthen their position in the market (Ghobakhloo, 2018; Wirtz, 2019). In

G. Schuh · C. Holper (✉) · L. Holst · V. Stich
Institute for Industrial Management (FIR) at RWTH Aachen University, Campus-Boulevard 55, 52074 Aachen, Germany
e-mail: Christian.Holper@fir.rwth-aachen.de

J. Meierhofer et al. (eds.), *Smart Services Summit*, Progress in IS,
https://doi.org/10.1007/978-3-031-36698-7_16

this course the technological potential of Industry 4.0 promises new opportunities to secure long-term growth and the ability to stay competitive in the market (Ghobakhloo, 2018). Consequently, numerous investments have been made in recent years. However, the predicted annual revenue increase for the manufacturing companies is not significantly visible so far (Koch et al., 2014). One reason for this is, that a breakthrough of service-driven business models—such as subscription business models—has not yet emerged (VDMA & McKinsey & Company, 2020). Subscription business models play a crucial role in the context of additional value generation as they have a high potential for the combination of value-added services and the physical product (Zhou et al., 2018). Accordingly, the focus of the business model is no longer on the product and its features, but on customer value and the provision of an overall solution. The main task is to shift value creation in the direction of the customer's value creation processes through a comprehensive subscription offering in conjunction with the corresponding revenue mechanisms (e.g., pay-per-use) (Grönroos & Ravald, 2011).

Despite its central importance for industrial companies, the implementation of subscription business models still represents a major challenge. Companies from the machine and plant engineering sector that have already started with the implementation of their subscription-based offerings are often confronted with a long and resource-intensive design process regarding the specific service offering (Richter & Tschandl, 2017). The industrial companies restart the design process for each new customer to fulfil individual customer needs (Nägele & Bading, 2011). However, the experience shows: customers' needs do not seem to vary as much as expected initially (Müller et al., 2015). Often benefit-based clusters can be identified, which help not only in creating a high value fit but also to standardize the provided service offerings.

Therefore, the aim of this paper is to identify specific customer types for subscription business models in the mechanical and plant engineering industry. To achieve this goal this paper will first identify relevant characteristics needed to describe industrial subscription customers and then define specific types based on their needs. While doing so the paper addresses the following research question: 'What specific customer types can be identified for the industrial subscription economy?'. A literature- and expert-based research methodology was used to identify the specific customer types.

2 Literature Review

2.1 *Industrial Subscription Business Models*

Originating in the media sector, different subscription-based offers have been developed over the last years, especially in the B2B segment ranging from consumable goods to software as a service (SaaS). In the recent past this business model approach has been adapted to the machine and plant engineering sector to use the benefits of industry 4.0 and to increase service sales. The underlying idea of providing a combined offering fit to the industrial customers' needs however is not new. Under the name of 'operator model' the concept has been discussed already by several authors (Freiling et al., 2004; Garrel et al., 2009; Spath & Demuß, 2006). Back then the associated risk transfer as well as the change of the value chain architecture have been focused. The main difference however is the fast development of information and communication technology. Due to the new technological possibilities the realistic analysis of the productive state of the machine is possible and allows the assumption of risks (Feldmann et al., 2021). Further new insights that are gained from the machine operation, can be used to improve processes and features attached to it. The combined solution offering allows to realize new value (Müller et al., 2018). These new value propositions in turn enable specific customer requirements to be met and accordingly create a high level of customer satisfaction.

3 Methodology

To answer the research question and create a systematic framework for the heterogeneous characteristics of the different customer types, the typology method is used in this paper. It allows to describe the characteristics of the different expressions and therefore provides a better understanding of the identified objects. In this paper we will follow the established approach by Welter (2006). The aim is to identify a systematic order for the investigated research area to make it more understandable by depicting relations among the characteristics and identifying individual types (Welter, 2006). Further the method allows to quickly visualize the field of all possible expressions and makes it therefore a practical tool desirable for industrial application.

The underlying framework of this paper includes all relevant characteristics and features as well as the customer types derived from it. To provide meaningful results, the following criteria must be met according to Schomburg (1980). First, each characteristic identified must have at least two distinct expressions, with no upper limit. Second, the characteristics must be all relevant for the explicit examination, so that there is a causal relationship between the purpose of the examination and the described characteristics. Third, the characteristics must be ascertainable, so that the objectivity and determinability is guaranteed in a sufficient quality. Finally, the characteristics must be differentiable to allow differentiation of the various types (Schomburg, 1980).

Based on this, Welter defines a five-step process for type creation (see Fig. 1).

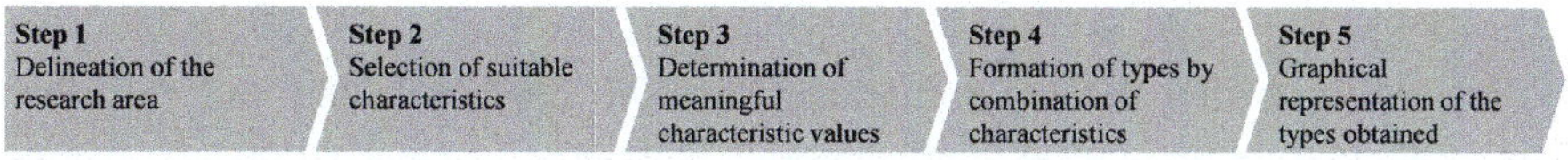

Fig. 1 Procedure for the creation of types (Welter, 2006)

The process starts with the delimitation and mental penetration of the field of research. Here the central object of investigation is determined with its typical characteristics. In the next step, the main characteristics are chosen that are suitable for describing the different types. Thereafter meaningful characteristic values are determined to describe the different expressions of each characteristic. In the fourth step specific combinations of characteristics are determined that form a certain type. Finally, the obtained results are then graphically presented.

The main results in form of the characteristics and the obtained types were derived logically from the experience of the authors and industry experts, considering current approaches in the literature and knowledge from operational practice. The interviewees came from different branches in the machine and plant engineering sector to increase the general applicability of the results. In addition, the customer types identified were validated afterwards with case studies and a discussion with experts from the industry round table 'Subscription Business Management'.

4 Results

4.1 Delineation of the Research Area

To determine a clear scope for this investigation, the research area needs to be delineated. The central focus of this study is on different customer groups in the context of a subscription-based offering. The term 'customer' is abstract and describes the demand side of the business relationship respectively the demanding company (Kittinger, 2011). When focusing on the customer, different observation levels involved in the buying process can be distinguished. According to the authors Webster and Wind there are four observation levels regarding the buying process (Webster & Wind, 1972). The characteristics and needs change depending on the level of observation.

The following paper will focus on the organizational level due to its comprehensive character. For this work it is important to look at the demanding company in total to consider the strategic goals and requirements that are to be met by using the subscription-based service offering.

4.2 Selection of Suitable Characteristics and Determination of Meaningful Characteristic Values

Within the next step the main characteristics are being selected and described. The focus here is on central characteristics that serve to clearly describe distinctive customer types. Hereafter, the characteristics are extended by corresponding characteristic values, describing the entire range of the corresponding characteristic.

The following framework consists of four main categories that describe the customers' characteristics. The categories are structured according to the idea of Shapiro and Bonoma (1984) in a hierarchical order, moving from a generic to a more specific perspective.

The first category deals with *generic* characteristics of the customer segment. The size of the company (Shapiro & Bonoma, 1984) plays an important role, as it can be an initial indicator of the extent to which services can be covered internally by the company itself. The evaluation is based on the company's turnover and ranges from micro-enterprises to large companies.

The second category focuses on *operational* characteristics that provide process-related information as well as insights about the strategy of the customer. The first characteristic identifies which processes of the customer are supported by the provider (Rüegg-Stürm, 2005). This classification makes it possible to better assess the need for the services used by the customer. A distinction is made between the customer's management, business, and support processes. The characteristic values help to analyze in which processes the customer requires support and consequently which needs arise as a result. As a further characteristic, the customer's strategy is analyzed as well as its current digital maturity level. In terms of strategy, the customer can focus on cost, quality, or innovation leadership (Porter, 2013). The digital maturity characteristic, on the other hand, describes the customer's level of technological maturity and its capabilities in relation to Industry 4.0. The characteristics are based on the industry 4.0 maturity index (Deutsche Akademie der Technikwissenschaften et al., 2020). Both characteristics expand the operational picture of the customer.

The third category focuses on *application-related* technological characteristics. The focus here is primarily on the components of the service offering that are obtained from the customer. A distinction is made between the physical product, the services, and the digital software. Regarding the three different types of service, a distinction is made as to whether the service is obtained from the provider itself, a competitor, or not at all. In terms of the physical product, it is also necessary to assess whether the supplier remains the owner of the product or whether ownership is transferred to the customer. Also, the customer's customization requirements are evaluated to assess the applicability of the existing service offering. Here, the range of services ranges from standardized to specifically developed services.

The fourth category covers *customer-specific* properties. Here, the customer's risks addressed by the service offering (Stoppel & Roth, 2016) and his general risk tolerance are assessed first. The corresponding risks relate to various business risks that arise during business activity. After that, the specific needs—according to their different benefit categories—are examined (Almquist et al., 2018). At first, the basic value characteristics are being observed. They represent criteria that can be seen as fundamental and accordingly essential for the customer's purchase decision. The value benefits on the next level address customers' performance-related needs, as for example an increase in sales or production quality. The characteristic values of the third group, the business-related needs, facilitate to do business. They provide objective values by improving its operational performance or productivity. The characteristic values on the organizational level address benefits, that concern the strategic

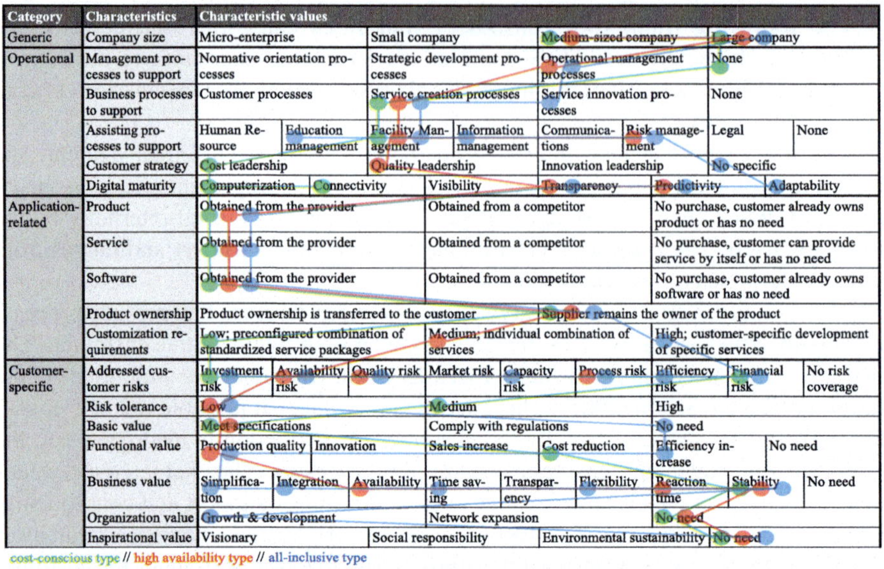

Fig. 2 Morphology of industrial customer types

orientation of the customer e.g. the expansion of its network. Finally, the inspirational value characteristics are concerned with the customer's visionary long-term goals to be achieved through business activity.

An overarching presentation of all relevant characteristics and their corresponding characteristic values can be found in the morphology presented below (see Fig. 2).

4.3 *Formation of Types by Combination of Characteristics*

In the last step specific customer types are being derived from the developed framework. The first customer type identified is the *cost-conscious type* (green). This type mainly concerns medium-sized to larger companies whose focus is strongly on achieving economic objectives. The companies are cost- or profit-driven and tend to be organized in profit/loss centers. This customer type often pursues a cost leader strategy. Regarding the range of services, the customer purchases product, service and software from the provider in order to obtain the most cost-efficient offer possible. However, to avoid high investment costs, the provider also remains the owner of the physical product. Key risks to be taken away from the customer are both the investment and the financial risk. The greatest benefit for the customer is a reduction in costs and a general increase in (financial) stability.

The second type identified is the *high availability type* (red). The focus of this customer type is the guaranteed availability of its machines. The customer's production processes require high machine reliability. Companies representing this type often pursue a quality leadership strategy and are therefore under high pressure to achieve corresponding production qualities. To ensure a high degree of availability, the companies require a high level of digital maturity. This type of customer obtains the entire range of services from the provider for the best possible monitoring of the production facilities. Also the range of services is partially aligned with the customer processes to be able to guarantee the best possible support. The most important risks assumed by the customer are availability, quality, and process risks. Overall, the general risk tolerance is very low. In addition to availability, key value benefits include production quality, fast response times and stability in the production process.

The third customer type is the *all-inclusive type* (blue). For this type of customer, the reduction in effort represents the greatest added value. The customer's goal is to have as little effort as possible with the corresponding machine application. In the best case, the customer can receive a ready-to-use solution that supports him in his business process. This type of customer is also prepared to pay accordingly. This type of customer usually refers to larger companies whose focus is clearly on their own value creation processes. The range of services provided by the supplier aims to take over all the customer's peripheral tasks associated with the machine availability. This extensive range of solutions makes it possible to transfer a large part of the customer risks associated with the machine application. The central needs of this type of customer, in addition to saving effort, are process simplification and long-term growth.

In addition to the description, the customer types obtained are being represented graphically below in the morphological box (see Fig. 2).

5 Conclusion and Outlook

Within this work distinct characteristics to describe industrial customers in the subscription economy were developed. Based on this descriptive characteristics, three different customer types were derived. The results aim to improve the process of customer-specific needs assessment by structuring and systematizing it and thus supporting companies in designing their fitting service offerings. Further the results aim to support future research by providing a structured overview of relevant customer-related characteristics. However, the results are also subject to limitations. Due to the high degree of novelty regarding subscription business models in the manufacturing industry the model presented requires continuous updating to stay relevant. This can be done by including new dimensions and characteristics. Further the model presented only refers to some sectors in the machine and plant engineering industry; additional interview partners from different subsectors are needed to provide a more generalizable picture.

References

Almquist, E., Cleghorn, J., & Sherer, L. (2018). The B2B elements of value: How to measure—and deliver—what business customers want. *Harvard Business Review*.

Deutsche Akademie der Technikwissenschaften, Forschungsinstitut für Rationalisierung e.V. an der RWTH Aachen, Technische Universität Darmstadt, Heinz Nixdorf Institut, Deutsches Forschungszentrum für Künstliche Intelligenz, & Fraunhofer-Institut für Materialfluss und Logistik. (2020). *Industrie 4.0 Maturity Index: Die digitale Transformation von Unternehmen gestalten*. UPDATE 2020 (Update 2020). *acatech Studie*. acatech - Deutsche Akademie der Technikwissenschaften.

Feldmann, S., Wienand, P., & Tonat, S. (07/2021). *X-as-a-Service transformation: From CapEx to OpEx. From products to solutions*.

Freiling, J., Buse, C., & Weißenfels, S. (2004). Strategische Grundfragen der Etablierung von Betreibermodellen. In H. Meier (Ed.), *Dienstleistungsorientierte Geschäftsmodelle im Maschinen- und Anlagenbau* (pp. 63–83). Springer Berlin Heidelberg. https://doi.org/10.1007/978-3-642-18578-6_4

Garrel, J. von, Dengler, T., & Seeger, J. (2009). Industrielle Betreibermodelle: One-Stop Services als erfolgreiches Konzept. In M. Schenk & C. M. Schlick (Eds.), *Industrielle Dienstleistungen und Internationalisierung* (Vol. 17, pp. 267–330). Gabler.

Ghobakhloo, M. (2018). The future of manufacturing industry: A strategic roadmap toward Industry 4.0. *Journal of Manufacturing Technology Management*, *29*(6), 910–936. https://doi.org/10.1108/JMTM-02-2018-0057

Grönroos, C., & Ravald, A. (2011). *Service as business logic: Implications for value creation and marketing* (Vol. 22). https://doi.org/10.1108/09564231111106893

Kittinger, A. (2011). Kunde. In A. Kittinger (Ed.), *Serviceorientierung und partnerschaftliches Handeln im B2B-Vertrieb* (pp. 11–16). Gabler. https://doi.org/10.1007/978-3-8349-6090-0_3

Koch, V., Kuge, S., Geissbauer, R., & Schrauf, S. (2014). Industrie 4.0. Chancen und Herausforderungen der vierten industriellen Revolution. *Strategy*. https://www.strategyand.pwc.com/de/de/industrie-teams/industriegueter/industrie-4-0-chancen/industrie-4-0.pdf

Müller, C., Nahr-Ettl, C., & Möller, R. (2015). Kundensegmentierung als Kernaufgabe. In J. Freiling & T. Kollmann (Eds.), *Entrepreneurial Marketing: Besonderheiten, Aufgaben und Lösungsansätze für Gründungsunternehmen* (2nd ed., pp. 103–120). Springer Gabler. https://doi.org/10.1007/978-3-658-05026-9_6

Müller, J. M., Buliga, O., & Voigt, K.-I. (2018). Fortune favors the prepared: How SMEs approach business model innovations in Industry 4.0. *Technological Forecasting and Social Change, 132*, 2–17. https://doi.org/10.1016/j.techfore.2017.12.019

Nägele, R., & Bading, A. (12/2011). *Industrielle Dienstleistungen: Erfolgsfaktor für die Maschinenbau-Branche der Region Stuttgart: Leitfaden zur Professionalisierung des Dienstleistungsportfolios*. Fraunhofer-Institut für Arbeitswissenschaft und Organisation (IAO).

Porter, M. E. (2013). *Wettbewerbsstrategie: Methoden zur Analyse von Branchen und Konkurrenten = (Competitive strategy)* (12. aktualisierte und erweiterte Ausgabe). Campus.

Richter, H. M., & Tschandl, M. (2017). Service Engineering—Neue Services erfolgreich gestalten und umsetzen. In M. Bruhn & K. Hadwich (Eds.), *Forum Dienstleistungsmanagement. Dienstleistungen 4.0* (pp. 157–184). Springer Gabler. https://doi.org/10.1007/978-3-658-17550-4_7

Rüegg-Stürm, J. (2005). *Das neue St. Galler Management-Modell: Grundkategorien einer integrierten Managementlehre ; der HSG-Ansatz* (8. Nachdr. d. 2. durchgesehenen u. korr. Aufl.). Haupt.

Schomburg, E. (1980). *Entwicklung eines betriebstypologischen Instrumentariums zur systematischen Ermittlung der Anforderungen an EDV-gestutzte Produktionsplanungs- und -steuerungssysteme im Maschinenbau* [Dissertation].

Shapiro, B. P., & Bonoma, T. V. (1984). How to segment industrial markets. *Harvard Business Review*(4).

Spath, D., & Demuß, L. (2006). Entwicklung hybrider Produkte—Gestaltung materieller und immaterieller Leistungsbündel. In H. -J. Bullinger, A. -W. Scheer, & K. Schneider (Eds.), *Service Engineering: Entwicklung und Gestaltung innovativer Dienstleistungen* (2nd ed., pp. 463–502). Springer Berlin Heidelberg. https://doi.org/10.1007/978-3-662-09871-4_19

Stoppel, E., & Roth, S. (2016). The conceptualization of pricing schemes: From product-centric to customer-centric value approaches. *Journal of Revenue and Pricing Management, 16*(1), 76–90. https://doi.org/10.1057/s41272-016-0053-1

VDMA, & McKinsey & Company (2020). *Kundenzentrierung als Chance für den digitalen Durchbruch. Was sich die Endkundenindustrien vom Maschinenbau bei digitalen Plattformen und Mehrwertdiensten wünschen.*

Webster, F. E., & Wind, Y. (1972). A General model for understanding organizational buying behavior. *Journal of Marketing, 36*(2), 12. https://doi.org/10.2307/1250972

Welter, M. (2006). Die Forschungsmethode der Typisierung: Charakteristika, Einsatzbereiche und praktische Anwendung. *WiSt*(2), 113–116.

Wirtz, B. W. (2019). Foundations of Digital Business Models. In B. W. Wirtz (Ed.), *Progress in IS. Digital Business Models* (pp. 1–15). Springer International Publishing. https://doi.org/10.1007/978-3-030-13005-3_1

Zhou, J., Li, P., Zhou, Y., Wang, B., Zang, J., & Meng, L. (2018). Toward new-generation intelligent manufacturing. *Engineering, 4*(1), 11–20. https://doi.org/10.1016/j.eng.2018.01.002

Business Models for Commercializing Predictive Maintenance in the Context of Industrial Sustainability

Jean Paul Potthoff, Christoph Tienken, and Thomas Friedli

Abstract As a digital technology, predictive maintenance (PdM) can contribute substantially to the necessary sustainable transformation of the manufacturing industry by maintaining assets more efficiently, increasing material and energy efficiency, and reducing associated waste. But as manufacturing companies struggle to implement sustainability, sustainable business models (SBM) for PdM might be a viable approach to systematically integrate sustainability. Hence the current PdM business models of manufacturing companies were analyzed by means of an exploratory case study. In a second step, elements of SBM that are already supported by today's PdM business models, as well as respective gaps to be addressed in the future, were identified. This study concludes that many elements of SBMs are already supported today, but the main emphasis lies purely on the economic benefits, and ecological and social benefits are currently disregarded. To develop SBMs for PdM, ecological and social aspects need to be systematically evaluated and integrated.

Keywords Predictive maintenance · Sustainability · Commercialization · Sustainable business model · Smart services

1 Introduction

Climate change is a pressing issue, for which manufacturing companies are at least partially responsible with their current way of business. To fulfill their pledges but also legal requirements in regard to sustainability, manufacturing companies should evaluate and implement digital technologies that can contribute to a sustainable transformation (Demartini et al., 2019). One of these technologies is predictive maintenance (PdM), with which machines can be maintained more efficiently, material and energy efficiency can be increased, and associated waste reduced (Waltersmann et al., 2021), thus enhancing sustainability for manufacturing companies (Burggräf et al., 2020). Many industrial companies are considering offering PdM as a service but are

J. P. Potthoff (✉) · C. Tienken · T. Friedli
Institute of Technology Management, University of St. Gallen, St. Gallen, Switzerland
e-mail: jeanpaul.potthoff@unisg.ch

J. Meierhofer et al. (eds.), *Smart Services Summit*, Progress in IS,
https://doi.org/10.1007/978-3-031-36698-7_17

unsure how to commercialize this software and data-based product in need of an IT infrastructure and support. Including sustainability in the resulting business model is a further challenge for most. Sustainable business models (SBM) are proposed as a possible solution in this context by the literature as a means to successfully commercialize a solution whilst also supporting sustainability (Bocken et al., 2013). Thus, the following research question is raised:

RQ: To what extent do current business models of PdM support SBMs? Which elements of the current business models need to be further developed to support SBMs?

To answer these questions, an exploratory case study approach was chosen. Drawing on data from nine interviews with providers of PdM (OEM or software developers for manufacturing companies), the current business models of PdM were analyzed. Relevant SBMs for PdM based on Bocken et al. (2014) were assessed and gaps regarding possible SBMs for PdM were identified.

2 Literature Review

PdM is a process that entails the "regular monitoring of the actual mechanical condition, operating efficacy, and other indicators [...] to provide the data required to ensure the maximum interval between repairs and minimize the number and costs of unscheduled outages created by machine-train failures" (Mobley, 2002). The resulting advantages are a reduction in production losses (Zoll et al., 2018), a reduction in maintenance efforts and costs, and a reduction of the probability of major failures (Darwanto et al., 2012). A growing body of literature on PdM being co-delivered to support sustainability is emerging (Abidi et al., 2022; Karuppiah et al., 2021). Generally, PdM can efficiently maintain machines, increase material and energy efficiency, and reduce associated waste (Waltersmann et al., 2021). This improved resource efficiency combined with extended product lifecycles helps to enhance sustainability (Burggräf et al., 2020) so that a direct link between the general value created by a PdM solution and sustainability can be made.

Even so, manufacturing companies generally have significant difficulties implementing sustainability into their business models (Bocken et al., 2014). They often focus on eco-design and eco-efficiency improvements, which help reduce energy consumption, resource intensity, emissions, and waste per unit of production, but are in sum not sufficient in facilitating the changes needed to ensure greater sustainability. In this context, Stubbs and Cocklin (2008) emphasize the relevance of SBMs, as they consider both a system and firm-level perspective whilst building on the triple bottom line. They also create a competitive advantage by contributing to a sustainable development and creating superior customer value (Luedeke-Freund, 2010). In this context, many different approaches to deliver sustainability through business models exist, which are based on different mechanisms and solutions (Bocken et al., 2014). Building on the value-based business model of Richardson (2008), Bocken et al. (2014) identified eight SBM archetypes, which describe the groupings of mechanisms and solutions. Of these archetypes, the following two SBM archetypes include elements that are relevant in the context of PdM:

(1) 'Maximise material and energy efficiency' (ME) is defined as doing more with fewer resources, generating less waste, emissions, and pollution. Examples are lean manufacturing and increased functionality, which all related to PdM.
(2) 'Encourage sufficiency' (ES) is defined as solutions that actively seek to reduce consumption and production. PdM can significantly increase assets lifetime, thus improving product longevity, requiring less purchase of assets, as well as educating operators on how to operate assets with less resource consumption.

Hence, both archetypes can serve as possible SBMs for PdM. Considering multiple archetypes to develop a SBM is recommended (Bocken et al., 2014). Hence, the current business model of PdM will be compared to both archetypes to identify gaps that could be further addressed by solution providers to co-deliver sustainability when offering PdM.

3 Methodology

An interview study was conducted for exploratory research, which is most suitable for unraveling complex detail (Eisenhardt & Graebner, 2007). This approach analyses individual observations and can transform these into generalizable patterns to derive new insights and theories. Thus, this approach allows the construction of new concepts that can be used to address the literary gaps.

A total of nine interviews, each with a different company, were conducted. All nine interviewees are currently involved or directly in charge of commercializing PdM solutions at their businesses. As this study seeks to understand the business model for PdM rather than the technologies themselves, interviewees with managerial rather than technical backgrounds were chosen. The interview guideline was structured according to the three business model dimensions after Richardson (2008). The transcribed interviews were all cross-case analyzed using the coding software *Atlas.ti*. Key insights were structured and compared to the elements of the relevant sustainable business models after Bocken et al. (2014) and visualized in the Figures in Sect. 4.

4 Results

The results are structured and presented following the business model dimension after Bocken et al. (2014), which were derived from Richardson (2008).

Value Proposition (VP) The primary value proposition (VP1) is the reduction of unscheduled (production) downtime, thus an increase in the Overall Equipment Effectiveness (OEE). An increase in the OEE allows customers to produce and sell more goods, hence VP1 is associated with an increase in revenue. The secondary value proposition (VP2) is the reduction of unscheduled maintenance, which is many times more expensive than planned and scheduled maintenance. Spare parts have to be express-delivered, and service technicians need to visit ad-hoc, all of which costs more than in the case of scheduled maintenance. Additionally, scheduled maintenance costs can also be reduced by moving away from preventive maintenance (VP3). An increase in asset utilization is achieved (VP4), giving customers more control and flexibility in their operations. These VPs are in line with maximizing asset productivity (increasing OEE) as well as resource efficiency (only conducting maintenance when necessary, thus supporting ME1 and ME2 (see Fig. 1). Waste in terms of unnecessary maintenance is reduced by moving away from preventive maintenance, the emissions are not addressed, thus only partially supporting ME3.

Asset life cycle can be extended through PdM, as major failures that would result in a complete loss of the asset are avoided and the overall maintenance of the asset, thus its condition, is improved (VP5). This in combination with the increase in asset utilization (VP4) results in improved asset durability and longevity, enabling slower asset replacement cycles, supporting ES1. ES2 is not addressed and hence is not supported (see Fig. 1).

Under consideration of these facts, PdM reduces unscheduled downtime, thus increasing revenue, and avoids unscheduled maintenance, thus reducing maintenance costs. These primary reasons are relevant from a financial perspective but do not contribute to sustainability. All further VP do contribute to sustainability but are communicated as "nice-to-have" side effects of PdM.

Value Creation and Delivery (VD) Identification of the value that can be generated via PdM for customers as well as the appropriate communication is key (VD1). The following capabilities are relevant in this context: (VD1.1) Detecting and understanding customer pain points and needs, e.g., understanding the impact of downtime on production and possible spillover effects as a consequence as well as understanding the mechanisms of maintenance costs. (VD1.2) Quantifying the value of PdM and providing a business case. As PdM is also a process that requires all involved parties to participate, the customer needs to be actively engaged, as only he can pro-

Current Business Model Value Proposition
VP1. Reduction of unscheduled downtime which is directly associated with increased revenue
VP2. Reduction of unexpected maintenance and its relatively high costs
VP3. Reduction of regular maintenance costs, e.g., better planning of maintenance activities and service missions
VP4. Increase asset utilization e.g., empower customers to further optimize their operations
VP5. Extend equipment life

SBM Archetypes Value Proposition	
ME1. Maximize material productivity	S
ME2. Maximize resource efficiency	S
ME3. Waste and emissions reduction	PS
Product and service solutions that seek to reduce demand-side consumption, e.g.	
ES1. Product durability & longevity, slow product replacement cycle	S
ES2. Educate on how to minimize usage impact	NS

Fig. 1 Comparison of the current PdM value propositions and the relevant SBM value propositions. S: Supported; PS: Partially Supported; NS: Not Supported

Current Business Model Value Creation & Delivery
VD1. Identifying and communicating the value for the customer
VD1.1 Detecting and understanding customer pain points and needs
VD1.2 Quantifying the value of PdM / Business Case first
VD2. Implementing PdM in a way that optimizes value creation and communicates clear results
VD2.1 Identification of point of interest/ critical assets and installation of sensors
VD2.2 Enable the integration of PdM with other systems
VD2.3 Communicating understandable results and initiating the maintenance process
VD3. Becoming a strategic partner in the value network

SBM Archetypes Value Creation & Delivery	
ME1. Activities and partnerships aimed at using fewer resources and generating little waste, emissions and pollution	PS
ME2. Focus is on product and manufacturing process innovation	S
ME3. New partnerships and value network reconfigurations to improve efficiencies	S
ES1. Ensuring activities, partners and customer relations are focused on consuming less, wasting less, and using products longer	PS
ES2. Product redesign for durability	PS
ES3. A shift in promotion and sales	PS
ES4. Supplier selection based on durability	PS

Fig. 2 Comparison of the current PdM aspects of value creation & delivery and the relevant aspects for SBM value creation

vide relevant domain knowledge of his operations and needs to implement necessary activities. A shift in promotion and sales is the consequence, away from a product-centric business towards a value-driven logic, though most companies are not able to manage this shift, thus partially supporting ES3 (see Fig. 2).

The following aspects contribute to an optimization of the value creation (VD2): Critical assets and points of interest need to be identified and appropriate sensors have to be installed. Only when doing so can PdM provide adequate value in the context of the first VP (reduction of unscheduled downtime). (VD2.2) The integration of PdM with other systems is necessary, e.g., triggering an appropriate sequence of activities to realize the theoretical PdM value. Production systems with model predictive control have the advantage that input from the PdM algorithm can be immediately processed, maximizing the PdM value. Both aspects focus on product and manufacturing process innovation, thus supporting ME2.

(VD2.3) Communication that is understood and respected by the customer and initiates the respective maintenance process. The customer needs to trust the PdM predictions, understand the necessary actions to be taken, and see the value in doing so. This communication ensures that the customer will use his assets more effectively, extend their lifetime and reduce waste (see VP). But as all of these activities happen under the premise to produce more in the end, which leads to an increase consumption of production inputs, ES1 is only partially supported. As there is currently no focus on reducing emissions or pollution, ME1 is also only partially supported (see Fig. 2). Overall, the activities conducted in VD2 lead to increased knowledge regarding the assets offered by the providers, on the basis of which the assets can get redesigned and supplier selection adjusted. But as it is unclear if all providers use the knowledge in such a way, ES2 and ES4 are only partially supported.

Lastly, providers of PdM try to establish a strategic partnership with their customers, as they can leverage the knowledge gained via PdM to generate further value

Current Business Model Value Capture
External value capture:
VC1. Value-based pricing:
▪ If the system can realize PdM improvements
▪ Value quantification based on customer financial and operational data
VC2. Else cost-or competition-based pricing
Internal value capture:
VC3. Lower service costs
VC4. Product improvement
VC5. Differentiation from the competition

SBM Archetypes Value Capture	
ME1. Costs are reduced through the optimized use of materials and reducing waste (PS)	PS
ME2. Compliance leading to increased profits and competitive pricing advantage (NS)	NS
ME3. Positive contribution through a minimized environmental footprint (NS)	NS
ES1. Profitability, customer loyalty, and increased market share realized from the provision of better products (durability)	S
ES2. Societal and environ. benefits captured	NS
ES3. Reuse across generations	PS

Fig. 3 Comparison of the current PdM value capture and the relevant SBM value capture

for both parties (VD3). Due to the data that is collected and analyzed as a result of implementing PdM, the providers can gain detailed insights into their customer's operations and current challenges. This knowledge can be used to further improve the PdM performance as well as customer operations. Such relationships are therefore used to educate the customer in regards to their operations as well as the improved maintenance of the assets. These strategic partnerships are new in their intensity and lead to improved efficiencies, supporting ME3.

Value Capture (VC) Value can be captured with the customer (external) as well as internally as a provider. Thus, this differentiation is applied (see Fig. 3).

External: Providers can establish value-based pricing (VC1) if the following requirements are met: (1) The production system is designed and operated in a way that the value of PdM can be realized, e.g., highly automated systems, where a prediction can immediately be processed and appropriate measures executed. (2) The value that is generated via PdM can be quantified. Value-based pricing incentivizes the provider to maximize the value for the customer. The value in question for this form of pricing is the increased revenue for customers via a reduction of unscheduled downtime as well as the reduction of costs due to unscheduled maintenance. Hence costs are reduced through the optimized use of assets, but the reduction of waste is not considered. Thus ME 1 is only partially supported (see Fig. 3). Cost- or competition-based pricing is otherwise established (VC2). Here, the provider has a predefined margin. This form of value capture does not have any effect on the SBMs.

Internal: Providers capture value internally in the following ways: (VC3) Lower service costs, e.g., the warranty phase, where PdM reduces the costs of warranty claims by enabling improved servicing of the machinery, also increasing the reputation. As more providers offer service level agreements (e.g., planning and carrying out maintenance activities, guaranteeing an OEE of X%), PdM helps to reduce service costs and improve the service quality, thus creating value for the customer and the provider, thus partially supporting ME1. (VC4) Providers of industrial equipment gain further knowledge regarding their products, which can be used to further improve the assets, but also reusing assets across generations (ES3). Understanding how the customers are running their operations can help to design, upgrade or maintain assets better. (VC5) Offering PdM is also a distinguishing factor from the competition, thus increasing or sustaining the current markets share. VC4 and VC5 thus support ES1 (see Fig. 3).

5 Conclusions and Recommendations

This study concludes that sustainability is currently of little relevance in the context of PdM. Providers prioritize the generation and capture of economic value (revenue increases and cost reductions) in their current business model but disregard possible ecological and social benefits, even though PdM can contribute significantly toward ecological sustainability as a digital technology. To help fulfill pledges and legal requirements regarding sustainability, providers should develop sustainable business models, of which relevant archetypes in the context of PdM were identified in this work.

As the current PdM business models already support many elements of these archetypes at least partially the following practical implications follow: (1) Providers should analyze how their PdM solution can provide sustainable value and put more emphasis on elements of SBM archetypes that are already supported. (2) The value proposition should emphasize contributions regarding material and energy efficiency, which can go hand in hand with a reduction of assets-, materials-, and other resource consumption, in addition to the (current) financial benefits. (3) The value creation and delivery should include activities that mitigate waste, emissions, and pollution as well as increase assets durability. New partnerships and network reconfigurations (e.g., combining domain knowledge of the provider and customer to improve operations) can improve sufficiencies, but require a constant dialogue and sharing of critical data. The reuse of assets across generations is already a success factor for some OEMs and should be considered by all.

Regarding the theoretical implications, this research gives an overview of the current commercialization of PdM based on the business model proposed by Richardson (2008). Building on the SBM archetypes of Bocken et al. (2014), gaps and opportunities for further research regarding sustainability in the context of PdM are identified. It is also the first work that discusses SBMs for PdM. More research is needed regarding the value capture of environmental and societal benefits that are realized by PdM, as this would significantly increase the attractiveness of SBMs. Further studies could extend the sample by focusing not only on providers of PdM but also on their customers and intermediaries. This study is solely based on qualitative research. To underline findings, future research should apply quantitative data to prove the influence of different business models with respect to economic and ecological value creation.

References

Abidi, M. H., Mohammed, M. K., & Alkhalefah, H. (2022). Predictive maintenance planning for industry 4.0 using machine learning for sustainable manufacturing. *Sustainability, 14*(6), 3387.

Bocken, N., Short, S., Rana, P., & Evans, S. (2013). A value mapping tool for sustainable business modelling. *Corporate Governance (Bingley), 13*(5), 482–497.

Bocken, N. M., Short, S. W., Rana, P., & Evans, S. (2014). A literature and practice review to develop sustainable business model archetypes. *Journal of Cleaner Production, 65*, 42–56.

Burggräf, P., Wagner, J., Dannapfel, M., Adlon, T., Saßmannshausen, T., Fölling, C., Föhlisch, N., Ohrndorf, D., & Minderjahn, R. (2020). Adaptive remanufacturing for lifecycle optimization of connected production resources—A literature review. In *Procedia CIRP* (Vol. 90, pp. 61–66). Elsevier B.V.

Darwanto, D., Hamdani, D., Hariyanto, D. D., & Karyawan, O. H. (2012). Partial discharge analysis for predictive maintenance of generator of geothermal power plant. *Proceedings of 2012 IEEE International Conference on Condition Monitoring and Diagnosis, CMD 2012* (pp. 740–743).

Demartini, M., Evans, S., & Tonelli, F. (2019). Digitalization technologies for industrial sustainability. In *Procedia Manufacturing* (Vol. 33, pp. 264–271). Elsevier B.V.

Eisenhardt, K. M., & Graebner, M. E. (2007). Theory Building from Cases: Opportunities and challenges. *Source: The Academy of Management Journal, 50*(1), 25–32.

Karuppiah, K., Sankaranarayanan, B., & Ali, S. M. (2021). On sustainable predictive maintenance: Exploration of key barriers using an integrated approach. *Sustainable Production and Consumption, 27*, 1537–1553.

Luedeke-Freund, F. (2010). Towards a conceptual framework of "Business Models for Sustainability". In *ERSCP-EMSU conference 2010* (Vol. 43, pp. 156–171).

Mobley, R. K. (2002). *An introduction to predictive maintenance*. Butterworth-Heinemann.

Richardson, J. (2008). The business model: an integrative framework for strategy execution. *Strategic Change, 17*(5–6), 133–144.

Stubbs, W., & Cocklin, C. (2008). Conceptualizing a "Sustainability Business Model". *21*(2), 103–127. http://dx.doi.org/10.1177/1086026608318042

Waltersmann, L., Kiemel, S., Stuhlsatz, J., Sauer, A., & Miehe, R. (2021). *Artificial intelligence applications for increasing resource efficiency in manufacturing companies-A comprehensive review.*

Zoll, M., Jack, D., & Vogt, M. W. (2018). Evaluation of predictive-maintenance-as-a-service business models in the internet of things. *2018 IEEE international conference on engineering, technology and innovation, ICE/ITMC 2018—proceedings.*

From Business Model to Contribution Model

Christoph H. Wecht, Elena Malakhatka, and Shaun West

Abstract A successful business is increasingly concerned with comprehending the challenges and opportunities associated with society's shift towards sustainability. However, current business model innovation fails to embrace the sustainability dimensions sufficiently. Many organizations lack a process that allows them generate entirely new and viable alternatives for business models. Sustainable development and its three-dimensional framework significantly impact most businesses' re-organization. The complexity of balancing the economic, environmental, and social dimensions of sustainable development signals the need for a new business model. In this conceptual paper, the authors discuss possible transitioning from the business model to the contribution model, where the value ecosystem and value exchange plays a defining role.

Keywords Business model · Contribution model · Sustainability · Value co-creation · Value ecosystem

1 Introduction

Most businesses strive to make the world a better place through their vision, leadership, culture, decision-making processes, management, business models, products, and services, regardless of the size and domain of the business. Today the current models for business do not fully embrace the sustainability dimensions. Rather, they

C. H. Wecht
Technology and Business, New Design University, St. Pölten, Austria
e-mail: christoph.wecht@ndu.ac.at

E. Malakhatka (✉)
Architecture and Civil Engineering, Chalmers University of Technology, Göteborg, Sweden
e-mail: elenamal@chalmers.se

S. West
School of Technology and Architecture, Lucerne University of Applied Sciences and Arts, Lucerne, Switzerland
e-mail: shaun.west@hslu.ch

J. Meierhofer et al. (eds.), *Smart Services Summit*, Progress in IS,
https://doi.org/10.1007/978-3-031-36698-7_18

remain based on 'profit' elements with negative consequences pushed 'off balance sheet'. Nevertheless, the negative consequences remain, and impact on the 'people' and 'planet' aspects of sustainability. To move towards a sustainable future, we must learn to capture value in a more equable manner that takes into account the unintended negative consequences, rather than focusing solely on the financial metrics associated with 'profit'. Therefore, we must re-consider value co-creation within the social, economic, ecological and technological systems and understand how business models contribute to people, plant and profit. Based on the fragmented nature of the existing models and approaches to balance different dimensions of sustainability within business models, we will use a literature review to support the development of a contribution model that considers the dynamic aspects of the ecosystems and the synergistic relationships between actors.

This article attempts to rethink the very purpose of the business model concept and its role in reviving our economy, environment and society through the contribution model framework. At the higher level, the framework consists of five areas, which are overwritten with the questions 'who,' 'what,' 'how,' 'why' and 'reasons'.

2 Towards Contribution Model Through Sustainability

There is a need for new business models that contribute to a sustainable development of the natural environment, society, and economy. This is where the field of Sustainable Business Model (SBM) research comes into play. Bocken et al. (2014), define sustainable business models as those that incorporate "a triple bottom line approach and consider a wide range of stakeholder interests, including environment and society" (p. 42). Geissdoerfer et al. (2018) mentions the "*creation of monetary and non-monetary value for a broad range of stakeholders and hold a long-term perspective*". Schaltegger et al. (2016) highlight in their definition of SBM the aspect of "*capturing economic value while maintaining or regenerating natural, social, and economic capital beyond its organizational boundaries*".

Most of the scholars in the SBM field propose that the business model should be sustainability orientated and follow an extended value creation to maintain a more holistic and systemic approach than in single bottom line business model development (Den Ouden, 2012). We agree that SBMs should define normative values, interests and goals related to multiple kinds of social, ecological and economic outcomes instead of placing the focal firm's economic profits in the center of consideration. Similar ideas were expressed by Lüdeke-Freund et al. (2017) with the proposal that the extended value creation principle demands that SBM development should contribute to generating not just value for single companies and their customers and shareholders but value for actors in both monetary and non-monetary terms. Thus, sustainability is becoming an increasingly powerful driver of business model transformation.

The preliminary answer to the first question was well described in Morone and Yilan (2020), where the authors describe a need for a fundamental paradigm shift in which business activities and consumption patterns can be aligned with environmental and social objectives. Answering the second question, we would like to bring few examples of economic transformation changes, happening in many industries right now and enabling a beginning of the paradigm shift towards more balanced sustainability approach in the business. One of such examples is a circular economy and its principles, which are well highlighted in different literature (Geissdoerfer, 2017; Planing, 2018). After defining the principles of a circular economy, where shared use, reuse and reduce are considered important for the overall value co-creation of the economy, Geissdoerfer et al. (2020) define circular business models "*as business models that are cycling, extending, intensifying, and/or dematerializing material and energy loops to reduce the resource inputs into and the waste and emission leakage out of an organizational system*" (p. 7). However, this perspective does not include most constituent business model aspects (e.g., value proposition or revenue model) but rather is a list of circularity measures. Another example of an attempt towards a more balanced sustainable business model is a sharing economy and collaborative consumption, which is often seen as a means of achieving sustainable development (Albinsson & Yasanthi Perera, 2012; Wecht et al., 2021). The sharing economy's popularity has increased in recent years as it presents an innovative consumption method that prioritizes access over ownership (Pérez-Pérez et al., 2021). Other examples came from industry and have moved into more altruistic models (i.e., Patagonia) where value is attributed to important non-financial (i.e., profit) based metrics.

We therefore conclude that the literature has not established a clearly defined connection between a sustainable business model and a significant contribution to sustainable development. However, we believe that important steps are being made in this direction, and one of the theoretical lenses that considers these transformational processes in the economy and business is service science, which often begins with a deep analysis of the value exchange and actors' network in a broader sense.

3 Re-thinking Value Co-creation in Sustainability-Oriented Service Systems

One of the key recommendations from this work is to understand value, the value creation processes, and the form of value. When Anderson et al. (2006) introduce value in a B2B environment, value is not only described in terms of financial values, but other less tangible forms are described. Raja et al. (2020) bring this to the current day and in line with Service-Dominant logic, (Vargo & Lusch, 2008) they confirm that two or more actors must interact within the service system to co-create value with and for the beneficiary. The value creation process is generally based on integrating tacit and explicit knowledge to deliver a value proposition. Other actors and equipment are often involved in the overall process. Value co-creation is not a single event and is based on multiple transactions, according to Grönroos (2011). The value created may be financial, tangible, or intangible, and additional tacit and explicit knowledge may be formed. Value itself is the net outcome of the benefits and negative aspects of the value proposition exchanged (Anderson et al., 2006).

The ecosystem resources limit the value co-creation, and changes in laws and behaviors may change the scale of the value (Kim et al., 2007; Story et al., 2017), as does the beneficiary. With changes to the environmental or intuitional arrangements, the form and quantity of the value may lead to value co-destruction (Frost et al., 2019). The business model of individual actors (at the ecosystem macro, meso and micro levels) explains in part the behaviors and motivations (Beirão et al., 2017) behind the multiple interactions that dynamically co-create value. This more profound understanding of the multiple interactions at different levels is essential within ecosystems (Vargo & Lusch, 2016). The quality of interactions, based on factors such as power and trust, can support or limit the value of co-creation (Fyrberg & Jüriado, 2009). A deeper understanding of ecosystems requires exploring value co-creation at various levels of aggregation (Vargo & Lusch, 2016).

Within the context of sustainability, value can be proxied based on output and categorized based on the dimensions of people, planet, and profit (Bergmans, 2006). The measurement of profit can be understood, although the metrics used may change. Within the ecosystem, firms opt to optimize their profitability often based on 'power' (Barney, 2018); trust, unless high within the ecosystem, can lead to a sub-optimal profit distribution (Williamson et al., 2012). Left alone, this can lead to unstable equilibria. Institutional arrangements and new norms (Imperial, 1999) can alleviate this unsustainable position, and new metrics are needed to aid the development of equable profit distributions. Planet costs need to be included within the accounting of value co-creation (Ansell et al., 2022). In effect, the impact on the planet needs to be accounted for within the ecosystem and the individual value propositions that define the interactions between actors (Grönroos, 2011; Vargo & Lusch, 2008). Without accounting for the damage, the negative impacts are not properly valued within the system. People are the individual actors at the meso-level and, in aggregate, represent the micro- and macro-levels. Proper measures and institutional arrangements can support the optimization of the ecosystem's value co-creation (Williamson et al., 2012). Yet, individuals are often inherently selfish and consider '*what's in it for themselves*', and some actors need to exhibit stewardship behaviors to reduce this selfish (or narcissist) tendency (Domínguez-Escrig et al., 2018).

4 Contribution Model Framework Development

Before starting to develop a new framework, we need to understand the drivers for different actors today. Management theory (Mintzberg, 2022; Rüegg-Stürm & Grand, 2019) states that businesses have a defined vison/purpose and from this they can define their strategy. The strategy leads to the business model and defines how the individuals within the firm interact together as well as with others outside of the firm. The business model (Gassmann et al., 2013) defines the "why, what, how and who" questions about how the firm interacts with others based on exchanges of value propositions with value based on 'value in context'. The firm is there, in its extreme, for the benefit of its shareholders (Porter, 2008).

A common vision and purpose have been shown to enable firms to act more sustainably. Yet, the prisoner's dilemma (Kuhn, 1997) confirms that without other control measures individuals and firms will continue to do what is best for themselves. Therefore, clear performance metrics are needed to help firms better understand their impact within the constraints of the People, Planet, Profit—3Ps (Fisk, 2010), this is needed to internalize what is today externalized and to prevent potential 'greenwashing'. In this case the business model needs to integrate the 3Ps into the performance measures. The new institutional arrangements, such as with financial reporting, need to be developed based around the 3Ps to support the visible reporting of sustainability.

The ecosystem perspective is important, within a sustainable business environment it is necessary to optimize the system (or local habitat). This way the value co-creation for the system beneficiaries can be maximized, which is important as otherwise we end up with a non-sum game and return to the prisoner's dilemma. Strong institutional arrangements are needed, and these must adapt to the local conditions. An approach to achieve this could be to integrate the 3P aspects into the business model. The institutional arrangements would have to report on the sustainability of the individual habitats within the overall ecosystems. It would also provide the opportunity to highly transitional ecosystem services necessary to support the changes. Within the traditional macro, meso and micro levels the reporting of the 3Ps needs to support the ecosystem to '*do the right thing*' and to make clear '*what's in it for me*', as only with active ecosystem integration can the best states come together (Malakhatka, 2021; West et al., 2018).

Our model expands the St. Gallen Business Model Navigator (Gassmann et al., 2013) approach in two distinct ways: it adds the area of 'reasons' and additional questions to the already existing areas. At the higher level, the framework consists of five areas, which are overwritten with the questions 'who,' 'what,' 'how,' 'why' and 'reasons'. Although the questions establish a connection to the previous approach, which focused exclusively on economic output, they are each regarded as a start for new comprehensive perspectives (Fig. 1). The extensions in detail are:

- *Who*: Stakeholder perspective, presented in a stakeholder matrix, instead of focusing on (paying) target customers.
- *What*: Value propositions structured along the triple bottom line instead of exclusive value propositions for paying customers.
- *How*: Joint value creation through networked cooperation, represented in a business ecosystem map, instead of the traditional sequential value chain perspective.
- *Why*: Financial and non-financial value capturing mechanisms presented in a comprehensive measuring card instead of the economic revenue model.

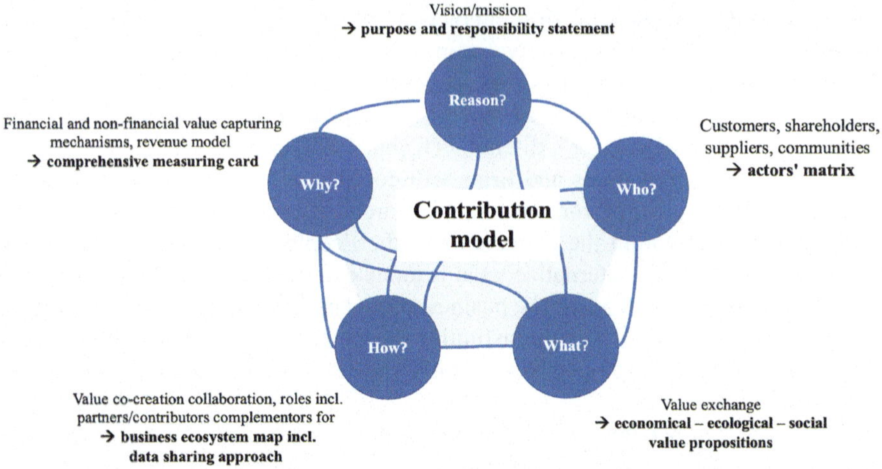

Fig. 1 Contribution model framework

– *Reason*: The purpose and responsibility statement of the organization is new in the framework in addition to the descriptions of vision and mission.

The business model describes how individual firms operate; by extending it to an ecosystem model, we start to understand how they all collaborate and contribute to different types of actors. Through extending the HSG business model, we better describe the individual, and by integrating it into an ecosystem, we can start to imagine new approaches to collaboration where we can account for both the 'goods' and the 'bads' and support sustainability.

For this reason, we use the ecosystem model to describe the web of transactions and relationships in the broader ecosystem; in the paper, we have shown a small habitat of the wider ecosystem. Within the ecosystem, competition, cooperation, and collaboration occur dynamically between various actors, including the beneficiaries. Equilibria can be formed, some more optimal than others, and sharing the good and bad from the 3Ps can support the dynamic optimization of value co-creation while minimizing the system level damage. Ecosystem services and institutional arrangements will be needed to facilitate and maintain the ecosystem or to support transitional periods. The contribution model ecosystem is presented below in Fig. 2 and shows different types of value exchange and contributions between multiple actors.

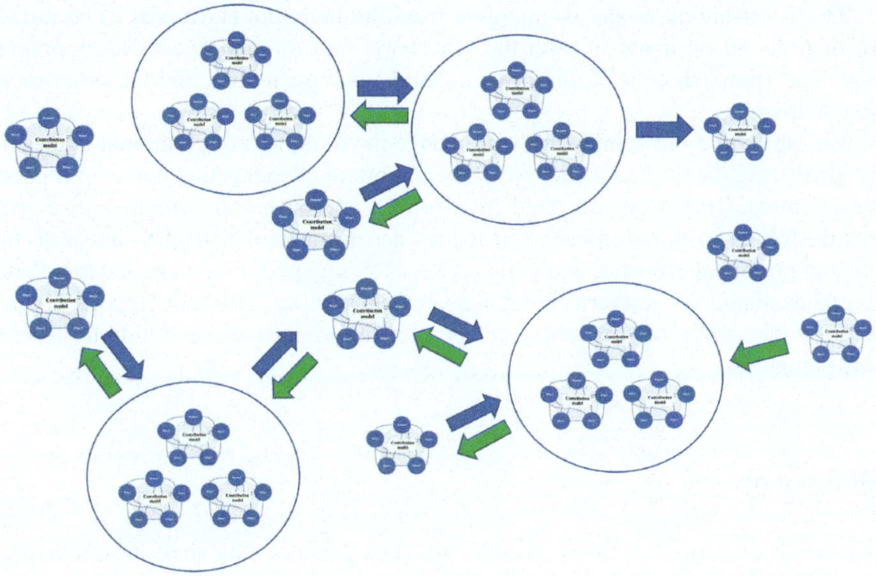

Fig. 2 Contribution model ecosystem showing exchanges

5 Reflections and Future Research

In this paper, we have taken different aspects of value co-creation from different perspectives and used them to help to pin down the basis of transactions and relationships within an ecosystem. The ecosystem we describe is not static but rather dynamic, allowing new and different forms of value creation through different resource integration. We also show that all elements are interlinked and that for sustainability to occur, the system level must be considered, rather than just the individuals within the system. Many existing business models are stand-alone and focus on the old world of tangible goods and financial gain. Value networks deal with expanded linear supply chains and fail to visualize the complexity of the value co-creation processes in data-enabled networks. Service science and the view of value co-creation, supported by institutional arrangements, provide an approach that can be highly academic and abstract. The method can be coupled with the ecosystem, where resources dynamically combine and re-combine, exchanging value propositions.

To capture the sustainability perspectives, we have adapted the traditional HSG business model to capture the complexity in real and dynamic ecosystems. The broader relationships between different contribution models are shown as an active ecosystem describing the exchanges between actors at two levels. The model could be extended to the meso, macro, and micro levels. Ecosystem services are not shown but could be added, along with the institutional arrangements supporting individual transactions between two or more actors.

This contribution model is emergent from the literature and needs to be tested. Testing should be made in both the real world and through agent-based simulation. The approach could follow two methods, one qualitative and the other more quantitative.

We aspire that our framework will change the thinking of operational managers by sensitizing them to do justice to all the important elements that make up modern management. Therefore, we build on a proven approach and complement it with practicable, practitioner-oriented structures and frameworks. Tool to integrate the most important approaches and perspectives affecting people/society and the planet (e.g., sustainability, circularity) into management practice. This increases the chance that a wider range of outcomes will be pursued alongside the traditional profit-oriented goals.

References

Anderson, J. C., Narus, J. A., & Van Rossum, W. (2006). Customer value propositions in business markets. *Harvard Business Review, 84*(3), 2–10.

Ansell, C., Sørensen, E., & Torfing, J. (2022). Co-creation for sustainability: SDG 17 as a lever of change. *Emerald Open Research, 4*(3), 3.

Albinsson, P. A., & Yasanthi Perera, B. (2012). Alternative marketplaces in the 21st century: Building community through sharing events. *Journal of consumer Behaviour, 11*(4).

Barney, J. B. (2018). Why resource-based theory's model of profit appropriation must incorporate a stakeholder perspective. *Strategic Management Journal, 39*(13), 3305–3325.

Beirão, G., Patrício, L., & Fisk, R. P. (2017). Value cocreation in service ecosystems: Investigating health care at the micro, meso, and macro levels. *Journal of Service Management.*

Bergmans, F. (2006). Integrating people, planet and profit. In Management models for corporate social responsibility (pp. 117–125). Springer, Berlin, Heidelberg

Bocken, N. M., Short, S. W., Rana, P., & Evans, S. (2014). A literature and practice review to develop sustainable business model archetypes. *Journal of Cleaner Production, 65*, 42–56.

Davis, J., Edgar, T., Porter, J., Bernaden, J., & Sarli, M. (2012). Smart manufacturing, manufacturing intelligence and demand-dynamic performance. *Computers & Chemical Engineering, 47*, 145–156.

Den Ouden, E. (2012). *Innovation design: Creating value for people, organizations and society* (p. 196). Springer.

Frost, R. B., Cheng, M., & Lyons, K. (2019). A multilayer framework for service system analysis. In *Handbook of service science.*

Gassmann, O., Frankenberger, K., & Csik, M. (2013). The St. Gallen business model navigator.

Geissdoerfer, M., Vladimirova, D., & Evans, S. (2018). Sustainable business model innovation: A review. *Journal of Cleaner Production, 198*, 401–416.

Geissdoerfer, M., Pieroni, M. P., Pigosso, D. C., & Soufani, K. (2020). Circular business models: A review. *Journal of Cleaner Production, 277*, 123741.

Grönroos, C. (2011). Value co-creation in service logic: A critical analysis. *Marketing Theory.*

Hahn, T., Figge, F., Pinkse, J., & Preuss, L. (2010). Trade-offs in corporate sustainability: You can't have your cake and eat it. *Business Strategy and the Environment, 19*(4), 217–229.

Hahn, A., Singh, R., Liu, C. C., & Chen, S. (2017, April). Smart contract-based campus demonstration of decentralized transactive energy auctions. IEEE.

Imperial, M. T. (1999). Institutional analysis and ecosystem-based management: The institutional analysis and development framework. *Environmental Management, 24*(4), 449–465.

Kaplan, R. S., & Norton, D. P. (2015). *Balanced scorecard success: The Kaplan-Norton collection (4 Books).* Harvard Business Review Press.

Kim, S.-H., Cohen, M. A., & Netessine, S. (2007). Performance contracting in after-sales service supply chains. *Management Science, 53* (12), 1843–1858.

Kramer, M. R., & Porter, M. (2011). *Creating shared value* (Vol. 17). FSG.

Kristensen, H. S., & Remmen, A. (2019). A framework for sustainable value propositions in product-service systems. *Journal of Cleaner Production, 223*, 25–35.

Kuhn, S. (1997). Prisoner's dilemma.

Lüdeke-Freund, F., Freudenreich, B., Schaltegger, S., Saviuc, I., & Stock, M. (2017). Sustainability-oriented business model assessment—A conceptual foundation. In *Analytics, innovation, and excellence-driven enterprise sustainability* (pp. 169–206).

Malakhatka, E., & Lundqvist, P. (2022). Actors' network analysis and Bi-Directional value exchange matrix development for living lab. In *Smart services summit* (pp. 139–147)

Mintzberg, H. (2022). An underlying theory for strategy, organization, and management: Bridging the gap between analysis and synthesis. *Strategic Management Review, 3*(1).

Morone, P., & Yilan, G. (2020). A paradigm shift in sustainability: From lines to circles. *Acta Innovations.*

Pérez-Pérez, C., Benito-Osorio, D., García-Moreno, S. M., & Martínez-Fernández, A. (2021). Is sharing a better alternative for the planet? *Sustainability, 13*(4), 1843.

Planing, P. (2018). Towards a circular economy-how business model innovation will help to make the shift. *International Journal of Business and Globalisation, 20*(1), 71–83.

Porter, M. E. (2008). The five competitive forces that shape strategy. *Harvard Business Review, 86*(1), 25–40.

Raja, J. Z., Frandsen, T., Kowalkowski, C., & Jarmatz, M. (2020). Learning to discover value: Value-based pricing and selling capabilities for services and solutions.

Rönnberg Sjödin, D., Parida, V., & Wincent, J. (2016). Value co-creation process of integrated product-services: Effect of role ambiguities and relational coping strategies.

Rüegg-Stürm, J., & Grand, S. (2019). *Das St. Galler management-modell: Management in einer komplexen Welt*. utb GmbH.

Schaltegger, S., Hansen, E. G., & Lüdeke-Freund, F. (2016). Business models for sustainability: Origins, present research, and future avenues. *Organization & Environment, 29*(1), 3–10.

Stefan, A., & Paul, L. (2008). Does it pay to be green? A systematic overview. *Academy of Management Perspectives, 22*(4), 45–62.

Stoll, O., Zou, W., Rodel, E., & West, S. (2021, November). Value co-creation in the context of digitally enabled product-service systems. In *Working conference.*

Story, V. M., Raddats, C., Burton, J., Zolkiewski, J., & Baines, T. (2017). Capabilities for advanced services: A multi-actor perspective. *Industrial Marketing Management, 60*, 54–68.

Stubbs, W., & Cocklin, C. (2008). Conceptualizing a "sustainability business model." *Organization & Environment, 21*(2), 103–127.

Vargo, S. L., & Lusch, R. F. (2008). From goods to service(s): Divergences and convergences of logics. *Industrial Marketing Management, 37*(3), 254–259.

Wecht, C. H., Cesinger, B., Vallaster, C., & Aleksic, N. (2021). Refocusing innovation management activities on sustainable circular business models-a framework and pathway for further research. In *ISPIM conference proceedings* (pp. 1–16).

West, S., Gaiardelli, P., Resta, B., & Kujawski, D. (2018). Co-creation of value in product-service systems through transforming data into knowledge. *IFAC-PapersOnLine, 51*(11), 1323–1328.

Williamson, P. J., & De Meyer, A. (2012). Ecosystem advantage: How to successfully harness the power of partners. *California Management Review, 55*(1), 24–46.

Innovating for a More Efficient World

Towards ANN Based Digital Twins of Ship Propulsion Systems

Dominik Rether, Martin Brutsche, Ioannis Sklias, and Markus Wenig

Abstract In shipping, the choice of the right routing and speed offers the opportunity to act more sustainably from both an economic and an ecological point of view. Reinforcement Learning (RL) agents could be suitable for this task. However, as a learning environment the agents require the most detailed, accurate, and fast representation of reality possible. This paper describes approaches to build such an environment using neural networks (NN) trained with both simulation and real-world data. It is shown that simple feed-forward networks can reproduce data created by 1D flow simulation sufficiently accurate. By examining the differences between simulation and measured data, the simulation could be improved. Since NNs trained with vessel data only are limited in their generality, approximating nets trained with simulation data to vessel data using Transfer Learning (TL) was investigated. Initial results for this approach show good quantitative results, but only in the data region where vessel and simulation data overlap. The paper provides an overview of the necessary steps towards Digital Twins for ship propulsion systems.

Keywords Machine learning · Digital twin · Ship propulsion systems

1 Introduction

Shipping is the backbone of the global economy. Despite a slight decline due to the pandemic, the mass of goods transported worldwide was 10.648 billion tons, 77.9% higher than in 2000 (United Nations Conference on Trade and Development, 2021), before China's accession to the World Trade Organization (WTO) and the resulting further integration of the Chinese market into global trade (Jungbluth, 2021). Events

D. Rether (✉)
Qnovi GmbH, Lothringer Str. 11, 70435 Stuttgart, Germany
e-mail: dominik.rether@qnovi.de

M. Brutsche · I. Sklias · M. Wenig
Winterthur Gas & Diesel Ltd., Schützenstrasse 3, 8401 Winterthur, Switzerland

J. Meierhofer et al. (eds.), *Smart Services Summit*, Progress in IS,
https://doi.org/10.1007/978-3-031-36698-7_19

such as the accident of the Ever Given in the Suez Canal in March 2021 show the economic and social importance of ocean shipping, as well as its irreplaceability for the global exchange of goods.

In principle, CO_2 emissions in ocean shipping per ton and kilometer traveled are the lowest compared to other means of transport (International Maritim Organization, 2020; Naturschutzbund Deutschland e.V., 2014). However, the other pollutant emissions that occur in particular during the combustion of heavy fuel oil are viewed critically (NABU, 2014). In recent decades, therefore, increasingly stringent emission regulations have been established for coastal areas as well as certain marine areas.

In addition to technological measures to reduce fuel consumption or pollutant emissions, there are also operational ones, such as weather-, current- and tide-optimized routing or real-time monitoring and optimization of ship parameters (Seum et al., 2011). To some extent, such measures are already being implemented, but mostly at the ship operator level (Seum et al., 2011) without giving to much technical details (CMA CGM Group, 2022; Hapag-Lloyd AG, 2022; MSC Group, 2022).

West et al. (2021) have shown in their article that the use of Digital Twins creates new opportunities for value co-creation in the sense of S-D logic. If we look at ocean shipping—in the following with a focus on the transport of goods—from the perspective of S-D Logic, it ultimately represents a transport and supply service. The goal is to flexibly transport as many goods as possible worldwide at low economic and ecological costs. Winterthur Gas & Diesel (WinGD) not only develops and designs marine propulsion systems for this service, but already offers a range of services, e.g. crew training for operation of propulsion systems, engine control updates for in-service engines to meet newer, more stringent regulatory requirements and 24 × 7 support for ship operators or their crews.

The development and manufacture of a ship engine represents only a fraction of the total cost of providing the utility service. Depending on methodology and system limits, about 55–65% of the total financial costs (LCC) of a vessel are incurred during vessel operation (Dinu & Ilie, 2015). If we look at CO_2 emissions, in some cases the percentage is even over 90% (Zhang et al., 2022). Thus, there is an opportunity for WinGD to identify ways to improve ship operations through the establishment of a smart service and to have meaningful operational measures implemented through appropriate operating instructions to the ship's crew. The charm of this service also lies in the fact that it can be applied to existing vessels or their propulsion systems. In connection with an average service life of cargo ships of approximately 25 years, this is therefore an interesting possibility for optimizing the vessels in operation.

The long voyage duration with all its imponderables, e.g. with regard to weather and sea state development, as well as the extensive logistics planning of ocean-going vessels, taking into account economic and regulatory constraints as well as dynamic port capacities, make a purely rule-based route optimization appear difficult and inappropriate. Karlsruhe Institute of Technology and WinGD have shown that Reinforcement Learning (RL) is suitable for solving such complex tasks (Moradi et al., 2022). They will show the CO_2 reduction potential through route optimization in their separate contribution to this Summit. The first row in Table 1 shows the

Table 1 Classification of smart-services according to West et al. (2021)

Service	Supported business functions	Service classification	Life-cycle	Environment	Capabilities	Decision-making horizon
Route optimization for vessels	Operations	Product life cycle services	MOL	Smart products	Optimization	Operational
Digital Twin of ship propulsion system	Development, operations	Product life cycle services	BOL, MOL	System of system	Monitoring, control	Tactical

classification of their service according to West et al. (2021). However, an important prerequisite for the successful learning of RL agents is an environment that fulfills the following criteria:

- Realistic reproduction of the system to be optimized.
- Fastest possible executability of the environment (several times faster than real-time).
- Easy parallelization of the environment.

Therefore, WinGD and Qnovi carried out a proof-of-concept, in which a simplified Digital Twin of a ship propulsion system is created using neural networks. However, the training of the nets is not carried out exclusively with measurement data, but also using simulation data. On one hand, this is based on the idea that simulations can be used to generate data as early as possible in the development process, before it is even possible to collect real measurement data. On the other hand, various boundary conditions can be better varied in the simulation, such as speed or ambient temperature. However, as soon as measurement data are available, they will be incorporated into the training, or existing networks will be adapted by means of Transfer Learning (TL). This Digital Twin is classified in the second row of Table 1.

2 Introduction of Project Data Sets

The proof-of-concept study was carried out for the WinGD 12X92DF. This engine with 12 cylinders and a bore of 92 cm has a base area of 230 m^2, is almost 16 m high and weighs 2140 t without operating fluids (WinGD, 2021). It is designed for dual-fuel operation, so it can basically burn Diesel or natural gas. In gas operation, CO_2 emissions are reduced compared with Diesel or heavy oil operation, and emissions of the other pollutants Sulphur Dioxides (SOx), NOx and particulate matters (PM) are mostly eliminated (Schneiter et al., 2019). From the dimensions of the engine, it is obvious that an experimental setup would be very expensive, which is why the development at WinGD had to be carried out purely virtually without a prototype (Schneiter et al., 2019). Accordingly, 1D flow models were also available, which can

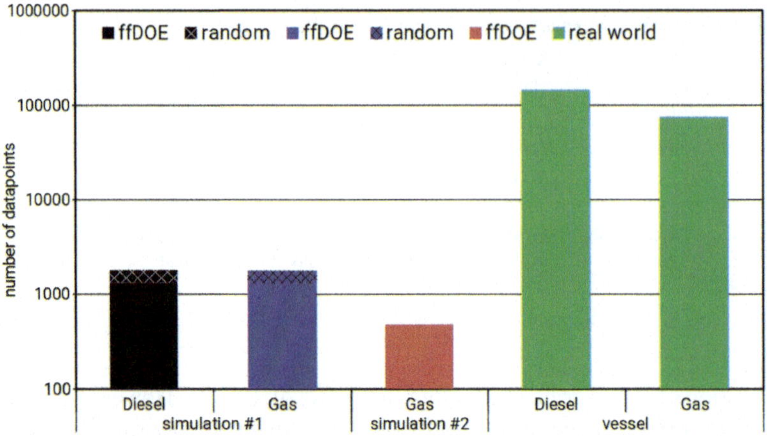

Fig. 1 Overview of available simulation and vessel data

predict physical quantities in the air path (such as turbine inlet temperature, scavenge receiver pressure or turbocharger speed) or the compression and peak pressure during combustion. The first ships with this engine have been in regular service since spring 2021, so that measured data with a resolution of 1/min are also available.

Figure 1 gives an overview of the simulation and measurement data used in the project. At project start, simulation data of a systematic speed/load variation along the propeller curve were already available. Part of the simulations were performed at five different ambient temperatures (ffDOE), while another 500 points were performed with again different speed/load combinations and random ambient temperatures (random). The vessel data covers an approximate five month period, with about 1/3 of the time in gas mode and 2/3 in Diesel mode.

In the following, only the gas mode will be discussed in more detail, for which an improved simulation data set (simulation #2) was available towards the end of the project. It contains 480 data points distributed over a slightly wider speed range, but was only created for three different ambient temperatures, see Fig. 2, upper left corner.

In general, it should be emphasized that the vessel data—although from only one ship—are about 100 times more extensive than the simulation data. Nevertheless, they cover a much smaller operational field as Fig. 2 shows.

3 Intialization of the Digital Twin Based on the Simulation Data

In comparable situations, smaller neural networks with three hidden layers and about 100 neurons per layer have proven successful (Milojevic et al., 2021, 2022). Since no fixed network layout is required in this project, the training was implemented as

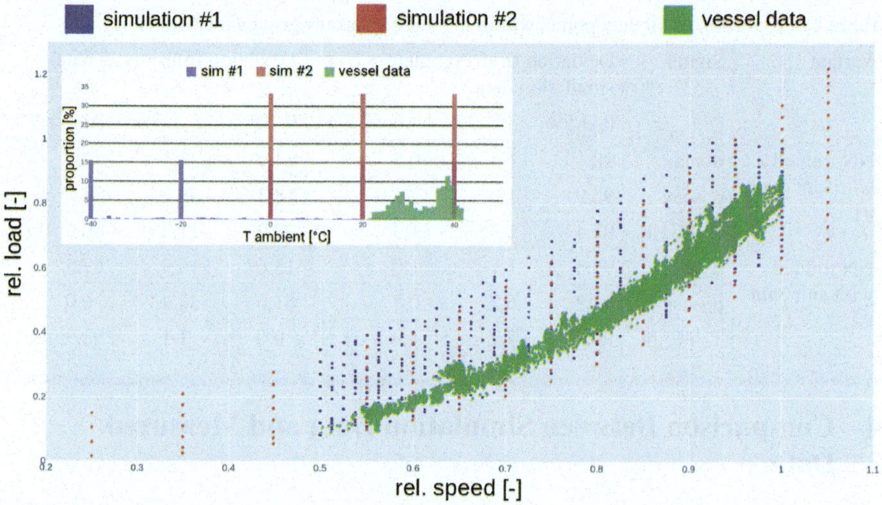

Fig. 2 Simulation #1 and # 2 and vessel data. (background: speed/load map, upper left corner: temperature distribution)

parallelized hyperparameter tuning, where the number of neurons and learning rate were randomly varied in predefined ranges. Since neural networks often show very unrealistic behavior as soon as they have to extrapolate, care should be taken when assembling the data that, the maximum and minimum values of all input and output signals are included in the training data set.

The further explanations are based on a division of training, validation and test data into 60%, 20% and 20%. Using just 20% for training and 20% for validation also gives acceptable results (not shown here) apart from areas of large gradients in the output signals. To ensure that overfitting was excluded, the histories of the training and validation loses were reviewed in a classical manner. However, with sufficient domain knowledge, the results in artificially generated data points can also be examined, for example, in the underrepresented temperature ranges. Initially, training was performed on the simulation data already available at the start of the project. After training, the neural nets were applied to all simulation and vessel data points. The three thermodynamic signals discussed here are an exemplary part of a detailed description of ship propulsion systems. The deviations of the signals predicted by the nets from the actual values were divided into three accuracy groups, see. line 2 in Table 2. For the most part, the nets can reproduce the simulation data very accurately, while there are some very large deviations in the vessel data, especially for the turbine inlet temperature (T_TC_in). This is not an error of the nets but can be explained by discrepancies between simulation and vessel data, which are explained in the following section. The second simulation data set available at the end of the project shows better quantitative agreement with the vessel data for scavange air pressure (p_scav), which then also applies to the networks trained with that simulation data.

Table 2 Proportion of all data points within the three accuracy groups for both NN

Variant	Signal	Deviation from complete simulation data			Deviation from vessel data		
		0–3.5%	3.5–7%	Above 7%	0–3.5%	3.5–7%	Above 7%
NN trained with sim data #1	p_max	98	1.5	0.5	24.9	23.1	52.0
	p_scav	96.9	1	2.1	18.3	28.2	53.5
	T_TC_in	97.4	1.3	1.3	7.9	11.2	80.9
NN trained with sim data #2	p_max	97.3	1.7	1.0	15.6	16.9	67.5
	p_scav	93.3	6.2	0.5	61.6	32.4	6.0
	T_TC_in	90.4	7.7	1.9	0.0	1.1	98.9

4 Comparison Between Simulation Data and Measured Data

In an ideal world, the simulation results and the vessel data would be congruent under the same boundary conditions, but this is often not the case. The error can be on both sides (simulation and measurement). Apart from model weakness, incorrect parameterization of simulation models or signal interference in measurements are both sources of error that should not be underestimated. Simulation models are often calibrated at certain reference points, i.e. the boundary conditions of the simulation are deliberately set to the reference points. However, the simulation data available in this project was generated for other purposes but should still be compared with the vessel data.

In principle, the Euclidean distance can be used to search in both data sets for pairs with similar values over all input variables of the Digital Twin. Since the value ranges of the various signals often differ—in this case by almost two powers of ten—the signals are normalized to their min–max values before calculating the Euclidean distance. This is to prevent that one of the input signals is too dominant. For the same reason, input signals of neural networks are generally normalized. As an example, p_scav will be examined for points with very similar boundary conditions. As Table 3 shows, the results of the second simulation are closer to the vessel measurements than those of the first simulation. However, as can be seen from Fig. 3, especially also the qualitative trends of the iso-lines is much more similar to the horizontal trends which can be seen in the vessel data. Especially when working with data generated with other objectives, it is therefore currently still important that people with the appropriate domain knowledge review the data. And since a generic model should be generated from the simulation data, it is of course very important to correctly reproduce the basic physical relationships.

Table 3 Deviation of the simulation data from the vessel data for similar boundary conditions

Variant	Number of similar data points	0–3.5% deviation	3.5–7% deviation	Above 7% deviation
sim data #1	59	23.7%	25.4%	50.9
sim data #2	30	43.3%	30%	26.7

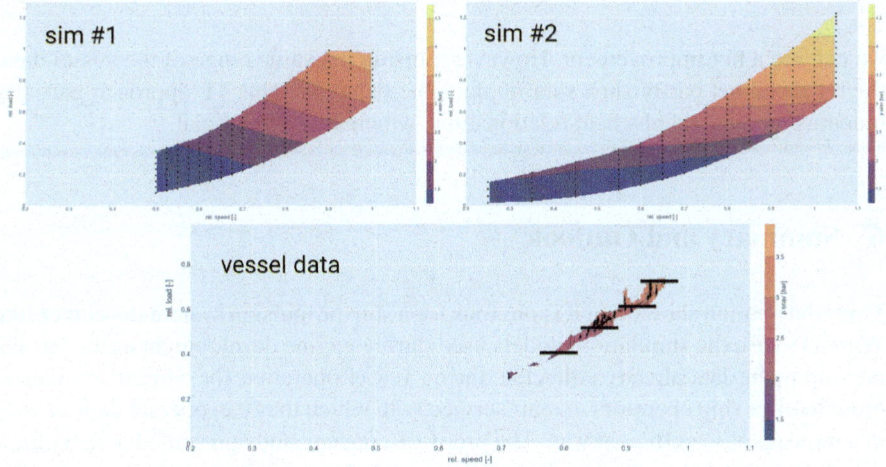

Fig. 3 p_scav over engine speed and load at constant ambient temperature (top left: sim #1, top right: sim #2, below: vessel data)

5 Adaptation of the Digital Twin to the Measurement Data

As the previous section has shown, despite the elimination of all human error, simplified simulation models will never represent the exact reality. Rather, they can be used to create generally valid, generic Digital Twins that are then adapted to the specific vessel. This is to be realized via Transfer Learning (TL). Usually, this method is used in image recognition, where Deep Neural Networks are trained in a generally valid way with a large computational effort—in order to learn the qualitative correlations—and then to be adapted to the specific use case with a much reduced effort in terms of training data and computational effort. Technically, this means that, for example, additional layers are added to the pre-trained net and then retrained with the data of the special use case.

In this study, a simplified variant was investigated in which the input layer and all hidden layers of the pre-trained net are blocked and only the output layer was retrained using the vessel data. Thereby, the qualitative behavior of the generic network shall be preserved to a large extent and only an adjustment of the value range shall be performed. Table 4 shows the accuracies achievable with this TL approach, using 20% of all vessel data as training data and another 20% for validation. At all signals

Table 4 Proportion of all data points within the three accuracy groups for NN after TL

Variant	Signal	Deviation from vessel data		
		0–3.5%	3.5–7%	Above 7%
NN trained with sim data #2 after TL with vessel data	p_max	70.7 (+55.1)	23.2 (+6.3)	6.1 (−61.4)
	p_scav	79.2 (+17.6)	14.7 (−17.7)	6.1 (+0.1)
	T_TC_in	90.7 (+90.7)	4.9 (+3.8)	4.4 (−94.5)

we can see a big improvement. However, outside the value range of the vessel data, the retrained net partly shows an implausible behavior. This TL approach partially unlearns the general physical relationships, which is not beneficial.

6 Summary and Outlook

This contribution shows that it is obvious for a ship propulsion system developer like WinGD to use the simulation models used during engine development as well as the measurement data already collected during vessel operation for current services in order to offer ship operators a smart service with which they can operate their vessels in a more resource-efficient way. The proof-of-concept study presented here outlines the steps necessary to create a Digital Twin of a ship propulsion system that can be used in the future as a virtual environment to train a Reinforcement Learning agent. The next project steps will include a more comprehensive investigation of alternative transfer learning concepts, as well as the establishment of a further automated process chain.

References

CMA CGM Group. (2022). Sustainable Development Report. CMA CGM S.A., Marseille. https://cmacgm-group.com/api/sites/default/files/2022-06/CMACGM_Rapport_2021_Web_UK.pdf

Dinu, O., & Ilie, A. M. (2015). Maritime vessel obsolescence, life cycle cost and design service life. *IOP Conference Series: Materials Science and Engineering, 95*, 012067. https://doi.org/10.1088/1757-899X/95/1/012067

Hapag-Lloyd AG. (2022). Nachhaltigkeitsbericht 2021. Hapag-Lloyd AG, Hamburg. https://www.hapag-lloyd.com/sustainability-report-2021/de/assets/downloads/HAPAG-LLOYD%20AG_Nachhaltigkeitsbericht%202021.pdf

International Maritim Organization (IMO). (2020). *Fourth IMO greenhouse gas study*. International Maritim Organization, London. https://wwwcdn.imo.org/localresources/en/OurWork/Environment/Documents/Fourth%20IMO%20GHG%20Study%202020%20-%20Full%20report%20and%20annexes.pdf

Jungbluth, C. (2021). *Challenge and opportunity: China inside the WTO and EU-China relations*. Bertelsmann Stiftung, Gütersloh. https://doi.org/10.11586/2021122. https://www.bertelsmannstiftung.de/fileadmin/files/user_upload/EU-China-Relations_Policy_Brief.pdf

Milojevic, S., Bodza, S., Bargende, M., Rether, D., & Grill, M. (2021). Next generation data based models powered by AI. In *9th international symposium on development methodology*, Wiesbaden. ISBN 978-3-9816971-7-9

Milojevic, S., Bodza, S., Cimniak, V., Angerbauer, M., Rether, D., Grill, M., & Bargende, M. (2022). *Data-driven modeling: an AI Toolchain for the powertrain development process*. SAE Technical Paper 2022-01-0158, WCX SAE World Congress Experience.

Moradi, M. H., Brutsche, M., Wenig, M., Wagner, U., & Koch, T. (2022). Marine route optimization using reinforcement learning approach to reduce fuel consumption and consequently minimize CO2 emissions. *Ocean Engineering, 259*, 111882. https://doi.org/10.1016/j.oceaneng.2022.111882

MSC Group. (2022). *Sustainability report*. Mediterranean Shipping Company SA, Genf. https://www.msc.com/-/media/files/sustainability/reports/msc_sustainability_report_2021.pdf

Naturschutzbund Deutschland e.V. (NABU). (2014). Luftschadstoffemissionen von Container schiffen. Naturschutzbund Deutschland e.V., Berlin. https://www.nabu.de/imperia/md/content/nabude/verkehr/140623-nabu-hintergrundpapier_containerschifftransporte.pdf

Schneiter, D., Goranov, S., Ott, M., & Printz, P. (2019). WinGD 12X92DF, the development of the most powerful Otto engine ever. In *CIMAC Congress 2019*, Paper 425, Vancouver.

Seum, S., Bahlke, C., Grasmeier, C., Gröger, J., Mottschall, M., & Schnegelsberg, S. (2011). Umweltschonender Schiffsbetrieb, PROSA Studie zum RAL Umweltzeichen UZ 110. Öko-Institut e.V., Berlin. https://www.oeko.de/oekodoc/1280/2011-412-de.pdf

United Nations Conference on Trade and Development (UNCTAD). (2021). *Review of maritime transport 2021*. ISBN: 978-92-1-113026-3. https://unctad.org/system/files/official-document/rmt2021_en_0.pdf

West, S., Stoll, O., Meierhofer, J., & Züst, S. (2021). Digital twin providing new opportunities for value co-creation through supporting decision-making. *Applied Sciences, 11*(9), 3750. https://doi.org/10.3390/app11093750

Winterthur Gas & Diesel Ltd. (2021). Marine Installation Manual X92DF

Zhang, Y., Chang, Y., Wang, C., Fung, J. C. H., & Lau, A. K. H. (2022). Life-cycle energy and environmental emissions of cargo ships. *Journal of Industrial Ecology*, 1–12.https://doi.org/10.1111/jiec.13293

Risk-Based Inspection for Smart Asset Management Decision Making on Major Rotating Equipment

Pascal Decoussemaeker, Sameer Vittal, and Shayan I. Ahmed

Abstract Asset management strategies for industrial process plants and utilities are based on inspection and test programs to periodically evaluate the progression of life consumption. A common shortcoming of such plans is that they are static and offer little flexibility for adaptation based on the actual operating experience. Data to improve these asset management plans is available from a variety of sources. Consolidating the data from these different sources is challenging but represents an opportunity for smart service providers to improve their offering and for operators to increase their production in a safe and reliable manner. Over the years, General Electric (GE) has been applying a variety of asset management technologies to improve power plant life and performance for their customers. These methods consider various inputs including equipment design knowledge, asset condition assessed from real-time monitoring and inspection data, and lessons learned from repairing, operating and maintaining thousands of assets globally. Risk-based inspection (RBI) represents a next generation of inspection and life management approaches. RBI is a consistent decision-making technique, which uses risk as a proxy for "asset condition," to manage the scope and frequency for the different activities in an asset life plan. This paper shows how GE has used RBI to perform risk-based health and life assessments for steam turbines and generators in power plants. The method has been used on several steam power plants around the world, to help operators of ageing steam plants make the right decisions to successfully navigate the global energy transition.

Keywords Asset management · Equipment life · Risk-based inspection (RBI) · Steam turbine · Generator · Smart services

---

P. Decoussemaeker (✉)
General Electric Company, Baden, Switzerland
e-mail: pascal.decoussemaeker@ge.com

S. Vittal
General Electric Company, Atlanta, USA

S. I. Ahmed
General Electric Company, Noida, India

J. Meierhofer et al. (eds.), *Smart Services Summit*, Progress in IS,
https://doi.org/10.1007/978-3-031-36698-7_20

1 Introduction

In asset intensive industries, asset life management strategies need to be defined to manage the life of the different assets and ensure trouble-free operation. For major components, these strategies are based on inspection and testing programs to evaluate if the progression of life consumption is not greater than expected. This is important to ensure that the investor will get enough life out of the equipment to ensure the expected return on his investment. As the equipment ages, there is also the question of when is the best time to renew a component? Premature renewal would lead to higher capital cost. However, if a renewal is delayed too long, the probability of an unplanned major breakdown increases.

For the implementation of these strategies, asset life management plans need to be determined to monitor and mitigate any risks associated with the asset life. Such a plan becomes smart when it is not static, but flexible and periodically updated as new data related to the asset condition becomes available.

2 Background: Risk and Failure Patterns

Everyone is confronted with risk assessments in daily life. Most decisions, like crossing a street, investing money or getting on an airplane, are based on a risk assessment. The risk of potential severe consequences is accepted if the probability is low enough (ASME, 2017). Risk is defined as the combination of the probability that a certain event occurs (PoF) and the severity of the consequence:

$$\text{Risk} = \text{Probability of Failure (PoF)} \times \text{Consequence of Failure (Severity)}$$

The PoF is an estimate of the uncertainty related to the equipment's continued resistance against the planned loading of the asset.

To define an asset life management plan, it is important to understand the specific characteristics of the failure modes that determine the life of the assets. Two main failure pattern categories can be distinguished, based on the behaviour of the failure probability (PoF) (see Fig. 1):

- **Age-related failure pattern**:

In this case, the asset will operate with a low probability of failure (PoF) for a certain time-period, defined as "Mid-Life Period" before it will reach a wear out phase where the risk of failure will increase rapidly.

- **Random failure pattern**:

For random failures, the conditional probability of failure does not increase with age. There is no relationship between age and reliability, therefore there is no preferential time to perform preventive maintenance activities.

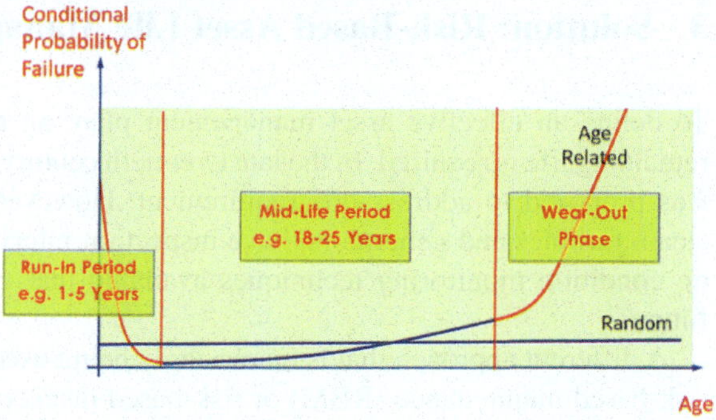

Fig. 1 Failure patterns

For both failure types, there may be an initial run-in period, with an increased risk of failure because of "initial settling issues and problems related to the re-assembly work". This type of failure is also referred to as "infant mortality".

The conditional probability of failure (Y axis in figure) is the probability that a piece of equipment fails during a certain time interval.

This paper covers the steam turbine and the generator of steam power plants. The dominant failure modes of these assets fall in the "age-related pattern" category.

For equipment of the same design, there can still be differences in the length of the mid-life period, because of differences in the material, operation profile and maintenance (see Fig. 2). The asset management strategy must therefore focus on:

- Identify any drivers (leading indicators) of accelerated life consumption at an early stage, in order to take measures to avoid premature degradation (see Fig. 2) and a reduction of the mid-life period, before it is too late.
- Adapt the strategy according to the phase in life. Once the asset is in the wear-out phase, the strategy should shift to short-term prediction of a possible failure. This will require increased inspection and monitoring.
- Because of this, it is important to predict the end of "mid-life" period and the start of the "wear-out phase" as accurately as possible.

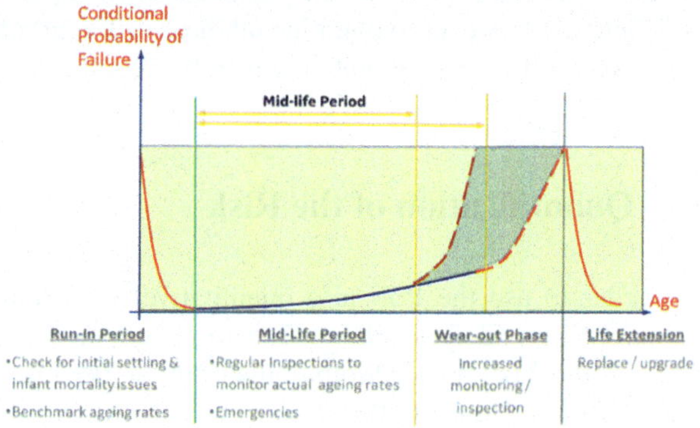

Fig. 2 Variation of the "mid-life period" and different focus areas throughout the life

3 Solution: Risk-Based Asset Life Management

To define an effective asset management plan, an accurate representation of the remaining life is required. In the late twentieth century, condition-based maintenance was proposed to address this requirement. However, power plant equipment like steam turbines and generators, have inspection intervals of 3–6 years and there are no condition monitoring techniques available that can predict 3–6 years into the future.

A different approach that is increasingly being used in various industries is called risk-based maintenance (RBM) or risk-based inspection (RBI). With this approach, the "risk of failure" is used as a proxy for the condition of the equipment. The benefit is that with a quantitative risk model, risk can be extrapolated over the life of the equipment and can therefore be used for lifetime management. A second advantage is that risk also includes the consequence of failure. Life management is therefore not just based on the potential occurrence of a failure, but also on the consequences, like for example loss of revenue or safety related issues.

RBI applies a structured approach and allows to integrate the available data for assessing, monitoring and mitigating risks related to the reliability of the assets. The results of the RBI assessment can then be used to make smart decisions to update the asset life management plan. Many variations of the RBI method exist and are currently being applied throughout different industries. Some best practices have been documented in "recommended practice" and "standard" documents by API (2016), APO (2019), ASME (2017), CEN (2008) and DNV (2010). Most of them are however focused on static equipment in the hydrocarbon or chemical industry (API, 2016, 2019), offshore oil and gas (DNV, 2010) or boilers (CEN, 2008). The method used by General Electric (GE) is derived from these different approaches but has been adapted, based on GE in-house experience and the specific challenges related to major rotating equipment on steam power plants.

The risk evaluation uses a large variety of data. There are the findings from the inspections and tests that have been performed over the life of the equipment, there is usually a history of many years of operational data stored in the site historian databases, condition monitoring results and the experience of the plant operators. The challenge is however to combine all these different inputs into a single, consolidated assessment that can generate the insights required for improved decision making.

4 Quantification of the Risk

In order to use the PoF in a quantitative risk evaluation, a modelling method is required. A widely used function to create such a quantitative model, from field data, is the 2 parameter Weibull distribution, because of its flexibility and simplicity (Davidson & Hunsley, 1994).

The PoF is calculated as follows with the Weibull formula:

$$P_f(t) = 1 - \exp\left[-\left(\frac{t}{\eta}\right)^{\beta}\right]$$

- Where Beta (β) is referred to as the shape factor and eta (η) as the scale parameter or characteristic life. Different values of β are used to model different shapes (see also Figs. 1 and 3).
- Weibull distributions with $\beta < 1$ have a failure rate that decreases with time, also known as infant mortality or early-life failures.
- Weibull distributions with β close to or equal to 1 have a constant failure rate, indicative of random failures.
- Weibull distributions with $\beta > 1$ have a failure rate that increases with time, also known as wear-out failures.

Since the methodology in the paper is limited to equipment with age related failure modes, the model needs to have a $\beta > 1$ for the RBI method to be applicable.

Fleet feedback from operational reporting and field service reports are used to determine the Weibull distributions for the risk assessment. In case insufficient data is available, a comparable proxy technology model of a different fleet can be used. The fleet average is modelled by the Weibull curve and the 90% confidence bounds are used to show the possible variation within the fleet (see Fig. 4).

To make a unit specific risk assessment, based on these fleet curves, a detailed ageing assessment is required. In the ageing assessment, unit specific operation data, inspection & test results and operator feedback are reviewed by means of a standard set of evaluation matrices (step 2 in Fig. 5). For each subsystem there is a dedicated matrix, based on its dominant failure modes. In this matrix, the leading indicators of life deterioration and possible symptoms are reviewed. The review is based on the inputs of inspection & test reports, answers to a questionnaire and an analysis of operational data for trends and anomalies. For every observation, there is a definition

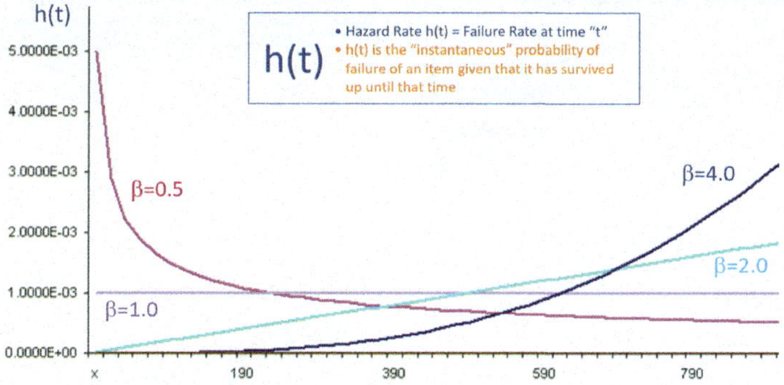

Fig. 3 Two parameter Weibull distribution model on graph that shows "conditional probability of failure" as a function of age

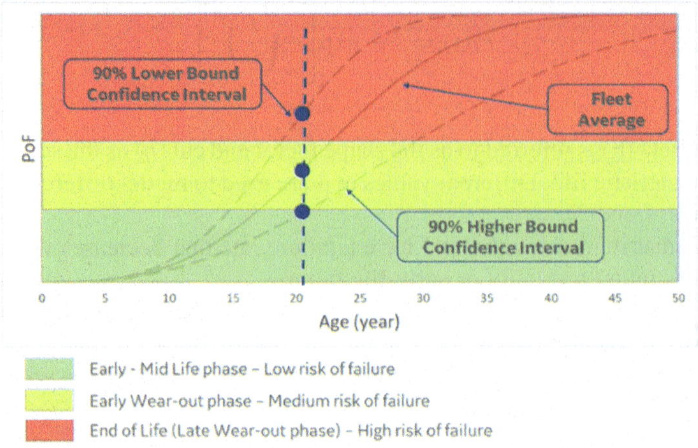

Fig. 4 PoF distribution

of what is "average", "below average" or "better than average". The worst-case evaluation item in each matrix is then used for the selection of the PoF curve (Fig. 4). The following questions need to be answered:

- Are there any leading indicators that indicate that the particular subsystem is ageing faster or slower than the average unit in the fleet?
- Are there any symptoms that indicate the presence of a developing failure?

Once it is clear which PoF curve (mean, higher or lower confidence bound) needs to be used, it is possible to do the risk assessment (step 3 in Fig. 5). Usually, an assessment is required for the present date, and for a certain date in the future (e.g., for a life extension of 10 years, it would mean 10 years in the future)? The PoF

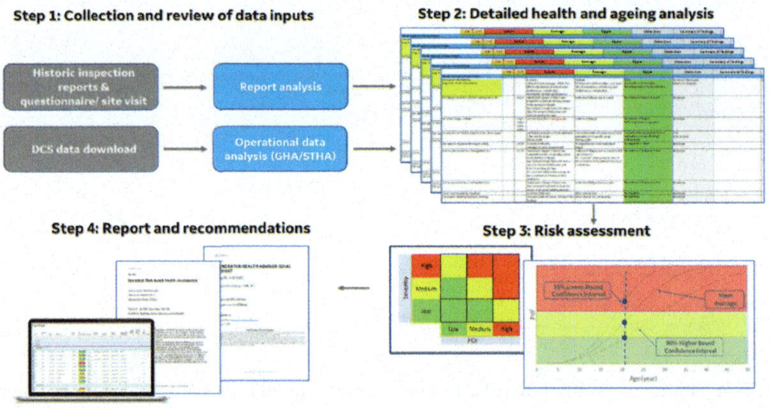

Fig. 5 Review process

needs to be taken from the PoF curve for the subcomponent, for the selected risk review dates (see Fig. 4). Once the PoF is determined, the risk can be calculated by combining the PoF with the severity of the potential consequences. After completion of the risk evaluation of the different subsystems, recommendations can be derived:

- If the risk is medium or high, and the PoF for today or the future date is in the "medium" zone (early wear-out phase, Fig. 4), additional monitoring or replacement needs to be recommended.
- If the PoF for today or the future date is in the "high" zone, a replacement needs to be recommended. Alternatively, a customer could set-up a business plan that accepts such an "end-of-life" event risk, through an emergency response plan.
- Additional recommendations can be made to mitigate the effect of leading indicators or to address any symptoms.

5 Results, Experience

General Electric (GE) has started applying this methodology on steam turbines and generators on several steam power plants around the world. Some examples of cases that were covered are listed below:

Example 1: Outage timing

Because of the Corona pandemic, it was not possible to perform a previously planned major outage on a water-cooled generator, because it was not possible to get the specialized staff into the country. The outage was therefore delayed by 3 years, and the plant operator wanted to know the risks related to this delay. Because some of the subsystems were getting into the "medium" risk zone and because the operational data analysis was showing some signs of rotor thermal sensitivity, it was recommended to monitor these critical areas more closely in order to get sufficient advance warning if any of those issues would start to deteriorate further.

Example 2: Life extension

A nuclear plant operator in the planning process for a 2 × 20 years life extension requested inputs for the generator. The method was used to either confirm some of the proposed life extension plans or advise to modify them. Specific fleet PoF curves were developed for this project, based on historic fleet experience.

6 Conclusion

Because of the on-going global energy transition, steam plant operators need to be flexible and adapt to changing requirements, rather than stick to what they have always done. Smart decisions, based on data, are needed related to the life management of the equipment. The wrong decisions may lead to high additional costs or risks, especially for the main equipment. It is therefore important for the decision makers to have the right insights available. GE has standardized its methodology to review the data from the different data sources in a systematic manner, and at the same time also uses fleet experience to create risk models. The outcome is a consolidated risk table that provides smart insights to help plant operators to define the next steps to best manage the life of their assets. Like this, data that was laying around, is now used to generate value, by providing insights for smarter decision making.

References

American Petroleum Institute (API). (2016). Risk-based inspection. API recommended practice 580 (3rd ed.).

American Petroleum Institute (API) (2019). Risk-based inspection methodology. API recommended practice 581 (3rd ed.). Addendum1.

American Society of Mechanical Engineers (ASME). (2017). Inspection planning using risk-based methods. ASME PCC-3-2017.

Davidson and Hunsley. (1994). The reliability of mechanical systems. IMechE Guides for the process industry, ISBN 10: 0852988818.

Det Norske Veritas (DNV). (2010). Risk based inspection of offshore topsides static mechanical equipment. Recommended practice DNV-RP-G101.

European Committee for Standardization (CEN). (2008). Risk-Based inspection and maintenance procedures for European Industry (RIMAP). CWA 15740:2008.

Machine Learning for Drone-Based Last Mile Delivery of Perishables

Selwyn Piramuthu

Abstract Last mile delivery of perishables is challenging in situations where necessary infrastructure to accomplish this task is lacking. However, the need for perishables, such as fresh blood for transfusion, does not disappear simply because of transportation-related issues. While last mile delivery has traditionally been done through ground-based modes, recent advances in drone-based technology calls for serious consideration of air-based delivery mode. In response to this call, current last mile delivery includes a mix of ground- and drone-based modes. In addition to the rest of the supply chain, the need to consider the last mile delivery aspect invariably renders the system more complex and requires the consideration of effective methods. Machine learning has been successfully used in complex systems. We therefore consider machine learning in various facets of last mile delivery scenarios with specific focus on classification, clustering, and optimization. We discuss possible applications of machine learning across these facets in drone-based last mile delivery of perishables.

Keywords Drones · Last mile delivery · Machine learning · Perishables · Sustainability

1 Introduction

Last mile delivery is a critical element in supply chains as packages need to be delivered to respective recipients regardless of local conditions such as rugged terrains or sparsely populated areas. In most cases, last mile delivery can be accomplished through any of several means such as bicycles, boats, drones, trucks, or vans. While bicycles, boats, trucks, and vans require appropriate conveyance infrastructure, drones are not restricted by such ground-based constraints. In recent years, drone use for last mile delivery has received increased attention as a viable option. The

S. Piramuthu (✉)
University of Florida, FL 32611-7169 Gainesville, USA
e-mail: selwyn@ufl.edu

J. Meierhofer et al. (eds.), *Smart Services Summit*, Progress in IS,
https://doi.org/10.1007/978-3-031-36698-7_21

existence of several companies in the last mile delivery space such as https://aerit.io/, https://www.flytrex.com/ and https://www.manna.aero/balbriggan attests to this fact. Other examples include Zipline and Wing that have been in operation for several years now and provide convincing use cases for drone-based last mile delivery. Since 2016, Zipline has delivered blood samples and medical supplies to rural hospitals and clinics in remote locations across Rwanda. Similarly, Wing has been delivering products from retailers such as Walgreens to residences in Christiansburg (Virginia) and Dallas-Fort Worth (Texas) areas.

Drone use for last mile delivery is generally touted as supporting environmental sustainability. However, from an environmental sustainability perspective, there are advantages and disadvantages to drone (vs. diesel truck) use in last mile delivery scenarios. In general, drones can help reduce air pollution in last mile delivery operations, with some caveats. Such last mile delivery through drones may necessitate additional warehouses and associated issues. Trucks with a fleet of drones can supplant such warehouses, where these trucks are parked at a launching site from where the drones are deployed to make last mile deliveries (Arishi, 2022). The lifecycle of drone batteries and how these batteries are or are not recycled must be factored in the assessment. As drones can only carry a few packages at a time, several trips are required to deliver all the packages destined for a given geographical area, unlike a truck which can carry several packages at once. Therefore, air pollution generated by last mile delivery vehicles must be calculated on a per package basis to ensure fair comparison. While the delivery of a truck-load of items to different destinations may not necessarily be appropriate for drone-based last mile delivery, situations that require individual deliveries such as urgent emergency medicine, blood samples to a nearby lab, or blood for transfusion (e.g., Rashidzadeh et al., 2021) to a single destination warrant serious consideration of drone-based delivery systems. In such situations where lightweight items are delivered over short distances, drone-based delivery can be quick and efficient with the avoidance of ground transportation while also using less (battery) power (vs. an electric delivery van). While taking the direct path from source to destination, such a setup also generally helps avoid land-based infrastructure and associated roadway congestion.

A variant of the last mile delivery is last yard delivery, especially in urban environments where the destination creates challenges due to unavailable space for delivering the items. For example, safe takeoff and landing pads or delivery spaces such as rooftops or necessary space even for dropping or parachuting packages. We do not consider the last yard delivery problem as the issues are disparate from that of last mile delivery.

While drone-based last mile delivery is relatively new in supply chains, machine learning for automated decision-making in various parts and aspects of supply chains is a very well-studied area (e.g., Liu et al., 2022; Piramuthu, 2005; Priore et al., 2019; Yan et al., 2022). We consider a few different facets of machine learning applications in supply chains that utilize classification, clustering, and optimization formats, with specific focus on drone-based last mile delivery. Unlike non-perishables, perishables face additional constraints on the ambient conditions they experience while in storage and transit. We study drone-based last mile delivery of perishables, possible decisions that need to be made in such environments, and associated constraints.

The rest of the paper is organized as follows. In Sect. 2, in addition to some background information, we identify and discuss related literature on machine learning for drone-based last mile delivery. In Sect. 3, we discuss some of the constraints faced by drone-based last mile delivery and then consider a few potential applications of machine learning across three different facets of drone-based last mile delivery. We conclude the paper with a brief discussion in Sect. 4.

2 Background and Related Literature

A report from Zipline (Fu, 2021) claims that their drone-based delivery generates respectively 99, 98, and 94% less carbon emissions compared to vans, cars, and electric vehicles. Similarly, a report on Wing (VT, 2020) estimated that drone-based delivery in a single US metropolitan area has the potential to reduce road use by up to 294 million miles per year and associated vehicle crashes by 580 while removing the equivalent of 25000 vehicles off the road, thereby reducing carbon emissions by up to 113900 tons per year. Rodrigues et al. (2022) studied the environmental impact of several last mile delivery methods and found that greenhouse gas emissions and energy consumption per parcel for drones were respectively 84 and 94% lower than those for diesel trucks. They observe that the speed of drone-based delivery is at the expense of environmental sustainability when a large number of packages are delivered to the same general area, while also noting that speed is important in situations such as when delivering an AED on time to treat a cardiac arrest patient or an EpiPen to save someone with nut allergy. These real-world examples illustrate the utility of drone-based last mile delivery.

Last mile delivery is generally accomplished through one of several variants that include ground vehicle-based, drone-based, and hybrid (e.g., Mora-Camino et al., 2021) ground vehicle-and-drone-based (Fig. 1). In all these scenarios, the items are picked up from a warehouse by either ground-based vehicles (e.g., vans, trucks) or drones. When ground-based vehicles are involved, the warehouse location can be farther away from the delivery area as distance is really not a hard constraint for such vehicles. In the ground vehicle-and-drone-based scenario, the warehouse can be equally farther away since the ground-based vehicle transports the items closer to the delivery area and then the drones take over. In the drone-based scenario, the warehouse is located relatively closer to the delivery area since drones have limited range. When the warehouse cannot be located close to the delivery area, ground-based vehicles are generally used to transport the items closer to the delivery area which is the same as the ground-vehicle-and-drone-based scenario.

While the literature on last mile delivery is extensive, there are relatively few that consider the incorporation of machine learning to make decisions. The utility of machine learning to support decision-making in a highly stochastic last mile delivery environment cannot be overstated simply because these are difficult problems to

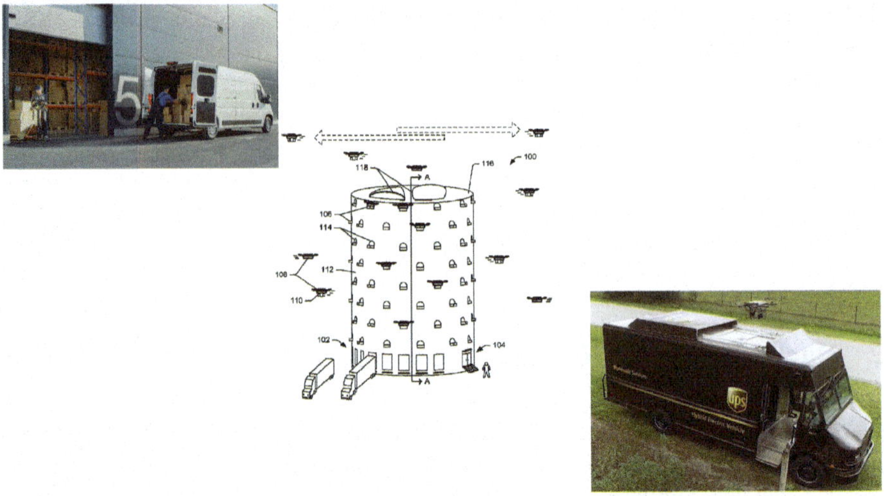

Fig. 1 Last mile delivery. *Sources* Coyote Logistics (2022), Curlander et al. (2017), Morris (2017)

address purely through deterministic means. Extant machine learning in drone-based last mile delivery literature considers the above-mentioned scenarios. For example, Chen et al. (2022) observe that ground-based vehicles can be slow due to urban traffic and drones have limited carrying capacity even though they are relatively fast, to develop a Deep Q-learning method to decide between drones and ground-based vehicles for last mile delivery. Bacanli et al. (2021) consider the problem of locating a warehouse between two sets of delivery neighborhood areas. The location of the charging and package pick-up station is determined through deep neural networks and support vector regressor based on package delivering times, package request frequency, and allowed package delays for each neighborhood. Their results show evidence of success for their method.

Sorbelli et al. (2022) study a ground-vehicle-and-drone-based scenario in which a truck carrying drones picks up the package at a warehouse and then releases the drones for delivery closer to the delivery site. They use drone energy cost, drone battery capacity, prioritized item deliveries, and time taken between launch and return of the drones to the truck to develop a heuristic algorithm. Similar to Sorbelli et al. (2022), Peng et al. (2019) consider cooperating ground-vehicle-and-drone-based scenario and propose a novel hybrid genetic algorithm for this setup.

de Araujo and Etemad (2021) use origin-destination parcel data from Canada Post that cover the Greater Toronto Area along with weather data to accurately predict drone-based last mile delivery durations. They use three types of convolutional neural networks and show that their results outperform those from classical machine learning algorithms. Sata et al. (2021) also consider a purely drone-based setup and model it as a sequential decision-making problem. They solve this problem with current and future item delivery demands through genetic algorithms as meta-heuristic function inside Trial-based Heuristic Tree Search (THTS).

From the brief review of extant literature on drone-based last mile delivery, we observe that the variant of choice depends on the distance between the warehouse and the overall package delivery area as well as the availability of a ground-based vehicle if the distance is beyond the drone range. It is also clear that machine learning has been studied in this context by several researchers, although the number of such studies is still quite small for each of the variants.

3 Machine Learning to Support Last Mile Delivery

As the feasibility of drone-based last mile delivery of items from warehouses becomes a reality, it is necessary to move beyond the basic ability to transport items from warehouse to end-user and to consider other aspects. There are several other aspects that are significant and need to be considered such as the physical aspect of the items to be delivered such as size, shape, and weight, privacy of the process in terms of the ability to protect the identity of the destination entity as well as the content of the package, and security of the entire system. While the physical aspects are addressed through physical means such as battery capacity, items that could be carried by the drone, and flying distance of the drone, security and privacy aspects are generally addressed through appropriate processes. Delivery drones also could face issues related to ethics, privacy, safety, and security. Ethical issues relate to what the drone must do in case of an imminent accident to ensure minimal harm to anyone as well as cause minimal damage to infrastructure. Privacy and safety issues relate to the payload's content as well as the recipient of that payload, especially when the payload is sensitive or private and its content or recipient is not meant to be divulged to an unauthorized party. Safety issues relate to collision with other objects or crash due to malfunction. Such an event could result in serious damage to property or harm to people especially when the drone payload is dangerous (e.g., caustic substance) or heavy.

As machine learning applications in last mile delivery of perishables gain traction, it is necessary to take stock of what has been done and to identify gaps that can be addressed. In general, machine learning can be used for classification (e.g., decision trees, feed-forward neural networks), clustering (e.g., centroid-based, density-based, distribution-based, hierarchical) and optimization (e.g., genetic algorithms) purposes. From a machine learning perspective, classification is the process of correctly placing data points into their categories, clustering is the process of clustering data points that are similar together in the same cluster while placing data points that are dissimilar to each other in different clusters, and optimization is the process of optimizing the objective functions given a set of constraints. We discuss possible applications of these facets in smart (e.g., Meierhofer et al., 2021, 2022; West et al., 2018) drone-based last mile delivery of perishables.

3.1 Classification

There are several decisions that could be modeled as a classification problem. Drones depend on their battery power to traverse their planned distance. When the battery cannot hold charge as expected or loses power faster than expected for any reason, the round trip to deliver a package to its destination and return to the warehouse or ground vehicle may not happen as planned. Some of the reasons for compromised battery power or lower drone range include low ambient temperature, wind, inclement weather conditions, and heavy package. Before a drone begins its trip, it needs to ensure that it will be able to return to a desired location after making its delivery. If estimates are that it cannot reasonably be assured of safe return, the trip needs to be scrapped. This is a classification problem with scrap or do not scrap the trip as the class variable (i.e., decision options). Independent variables could include delivery urgency (i.e., if a hot job), visibility, rain/no-rain, ambient temperature, wind speed and direction, payload weight, and payload physical dimensions. The decision on what heuristic to use to schedule the delivery of items is a classification problem. Based on a snapshot of the system at any point in time, the appropriate heuristic to use (e.g., send the package that is closest in geographical distance first, send the items that are the most sensitive to temperature variance first). To our knowledge, none of the published research in this general area models these as classification problems.

3.2 Clustering

Clustering can be used to determine which delivery addresses belong to a delivery area. With knowledge of the coverage of a given delivery area, a set of drones can then be assigned to this delivery area from a warehouse or a ground-based vehicle. The clusters themselves can be formed based on the distance between each delivery address as well as other factors that include the package characteristics (e.g., weight, physical dimensions), package delivery urgency, and availability of appropriate drones that can be summoned to deliver such packages on time. For example, if a delivery address does not have assigned drones that can deliver a given packet due to its weight, that package can be assigned to a delivery area with an appropriate drone that can accommodate that weight and distance range from the warehouse or ground-based vehicle.

Another application is the generation of clusters of delivery jobs for each drone, taking into account the constraints associated with the carrying capacity (e.g., weight, number of packets) and the sequence delivery requirements for a given drone.

3.3 *Optimization*

There are several optimization problems in the context of drone-based last mile delivery. For example, the delivery sequence for each drone can be optimized based on a performance metric such as minimizing the average customer wait time, whereby parcels are arranged in increasing order of required time to deliver and are delivered as per this order. The assignment of drones to different delivery areas can also be optimized based on the number and type of packages as well as the number and type of drones that are assigned for each delivery area. Routing can be optimized for each drone so that it takes the shortest route from source to destination and back—this needs to be done for all drones assigned to a delivery area to reap the benefits of optimal routing.

4 Discussion

In the United States, the Federal Aviation Administration (FAA) developed a set of restrictions on non-recreational drones that are used for package delivery (FAA, 2016). While some of these such as maximum altitude (400 feet) and maximum speed (100 mph) do not affect most drones, others such as avoidance of populated areas, maintaining at least 400 feet distance from any building structure, not operating when visibility is less than 3 miles, and VLOS (visual line of sight) requirement restrict drone operations for last mile delivery. Therefore, drone use for last mile delivery needs to take local regulations into consideration.

We considered drone-based last mile delivery of perishables and identified a few possible ways in which machine learning can be used in this domain. Under appropriate circumstances, drone-based delivery systems create value (e.g., Meierhofer et al., 2021, 2022, West et al., 2017, 2020) when they supplant existing ground vehicle-based delivery systems. While drone use is not always the best from an environmental sustainability perspective, there are cases in which drone use is not only effective and efficient, but also generates lower carbon footprint. In these cases, where one or a few small lightweight items are urgently needed not far from a warehouse that stocks these items, delivering such items even through ground-based electric vehicles results in about 30 times (Fu, 2021) the carbon footprint as compared to drone-based delivery. Such scenarios require the best possible choices at its decision points. We discuss the use of machine learning for such decisions. From our review of extant literature in this general area, we observe that the number of research publications on machine learning use for drone-based last mile delivery of perishables is rather limited. Given the significance of drone-based last mile delivery, the hope is that the potential of machine learning is fully utilized to effectively serve those in need in a timely manner while supporting environmental sustainability.

References

Arishi, A. (2022). A hybrid learning model for the vehicle routing problem with drones in the era of industry 4.0. *Proceedings of the 18th annual symposium on graduate research and scholarly projects*. Wichita State University. https://soar.wichita.edu/handle/10057/23486

Bacanli, S. S., Cimen, F., Elgeldawi, E., & Turgut, D. (2021). Placement of package delivery center for UAVs with machine learning. *IEEE Global Communications Conference (GLOBECOM)*, 1–6. https://doi.org/10.1109/GLOBECOM46510.2021.9685951

Chen, X., Ulmer, M. W., & Thomas, B. W. (2022). Deep Q-learning for same-day delivery with vehicles and drones. *European Journal on Operational Research, 298*(3), 939–952.

Coyote Logistics. (2022). https://coyote.com/ship-freight/final-mile/

Curlander, J. C., Gilboa-Amir, A., Kisser, L. M., Koch, R. A., & Welsh, R. D. (2017). *Multi-level fulfillment center for unmanned aerial vehicles*. https://patentimages.storage.googleapis.com/cb/07/c0/3cd597a07d5198/US9777502.pdf

de Araujo, A. C., & Etemad, A. (2021). End-to-end prediction of parcel delivery time with deep learning for smart-city applications. *IEEE Internet of Things Journal, 8*(23), 17043–17056.

FAA. (2016). *Summary of small unmanned aircraft rule (Part 107)*. 21 June. https://www.faa.gov/uas/media/Part_107_Summary.pdf

Fu, E. (2021). *The sustainability of zipline's autonomous aerial logistics*. Flyzipline.com

Kitjacharoenchai, P. (2020). Optimization models and analysis of truck-drone hybrid routing for last mile delivery. *Ph.D. (Industrial Engineering) dissertation, Purdue University*. https://hammer.purdue.edu/ndownloader/files/22311804/preview

Liu, N., Bouzembrak, Y., van den Bulk, L.M., Gavai, A., van den Heuvel, L.J., & Marvin, H.J.P. (2022). Automated food safety early warning system in the dairy supply chain using machine learning. *Food Control, 136*, 108872

Meierhofer, J., Benedech, R.A., Schweiger, L., Barbieri, C., & Rapaccini, M. (2022). Quantitative modelling of the value of data for manufacturing SMEs in smart service provision. In *12th international conference on exploring service science*. https://doi.org/10.1051/itmconf/20224104001

Meierhofer, J., Heitz, C., & Hannich, F. (2021). Optimizing service value creation with smart, connected products. *Proceedings of the Naples forum on service*. https://naplesforumonservice.com/wp-content/uploads/2021/09/meierhofer-heitz-hannich-optimizing-service-value-creation.pdf

Meierhofer, J., Heitz, C.: Service customization: Optimizing value creation and capture by designing the customer journey. In *Proceedings of the 8th swiss conference on data science (SDS)* (pp. 32–38)

Meierhofer, J., Kugler, P., Vogt, H., Dobler, M., Benedech, R. A., Strittmatter, M., & Treiterer, M. (2022). Improving service value creation for manufacturing SMEs by overcoming data sharing hurdles in ecosystems. In *Spring servitization conference (SSC)* (pp. 86–94)

Mora-Camino, F., Raoul, J.-L., Lamiscarre, B., Mykoniatis, G. (2022). Last mile delivery model for sustainable development. In *9th international conference on experiments/process/system modeling/simulation/optimization*. https://hal-enac.archives-ouvertes.fr/hal-03313087/document

Morris, C. (2017). UPS tests drone launched from EV for last-mile delivery. *Charged—Electric Vehicles Magazine*. https://chargedevs.com/newswire/ups-tests-drone-launched-from-ev-for-last-mile-delivery/

Peng, K., Du, J., Lu, F., Sun, Q., Dong, Y., Zhou, P., & Hu, M. (2019). A hybrid genetic algorithm on routing and scheduling for vehicle-assisted multi-drone parcel delivery. *IEEE Access, 7*, 49191–49200.

Piramuthu, S. (2005). Machine learning for dynamic multi-product supply chain formation. *Expert Systems with Applications, 29*(4), 985–990.

Priore, P., Ponte, B., Rosillo, R., & de la Fuente, D. (2019). Applying machine learning to the dynamic selection of replenishment policies in fast-changing supply chain environments. *International Journal of Production Research, 57*(11), 3663–3677.

Rashidzadeh, E., Hadji Molana, S. M., Soltani, R., & Hafezalkotob, A. (2021). Assessing the sustainability of using drone technology for last-mile delivery in a blood supply chain. *Journal of Modelling in Management, 16*(4), 1376–1402.

Rodrigues, T. A., Patrikar, J., Oliveira, N. L., Matthews, H. S., Scherer, S., & Samaras, C. (2022). Drone flight data reveal energy and greenhouse gas emissions savings for very small package delivery. *Patterns*, *3*, 100569. https://doi.org/10.1016/j.patter.2022.100569

Sata, B., Lacan, J., & Chanel, C. P. C. (2021). A GA-guided trial-based heuristic tree search approach for multi-agent package delivery planning. *Scheduling and planning applications workshop (SPARK)—International conference on automated planning and scheduling (ICAPS)*

Sorbelli, F. B., Corò, F., Das, S. K., Palazzetti, L., & Pinotti, C. M. (2022). Greedy algorithms for scheduling package delivery with multiple drones. In *23rd international conference on distributed computing and networking (ICDCN)* (pp. 31–39)

VT. (2020). *Drones deliver time savings to customers, new revenue to businesses*. https://vtx.vt.edu/articles/2020/09/ictas-dronedeliveryeconomics.html

West, S., Meierhofer, J., Stoll, O., & Schweiger, L. (2020). Value propositions enabled by digital twins in the context of servitization. *Proceedings of the spring servitization conference : advanced services for sustainability and growth* (pp. 152–160)

West, S., Mueller-Csernetzky, P., Kuenzli, M., Huonder, M., & Granata, T. (2018). Ecosystems innovation for smart connected services. In *25 th International EurOMA conference*

West, S., Granata, T., Kuenzli, M., Ouertani, Z., & Ganz, G. (2017). Identification of ecosystems actors and their behaviours in manufacturing and services. *Frontiers in Service*, 22–25

Yan, Y., Gupta, S., Licsandru, T. C., Shoefer, K. (2022). Integrating machine learning, modularity and supply chain integration for Branding 4.0. *Industrial Marketing Management*, *104*, 136–149

Reducing Service Trips and Its Effects on Service Delivery and Customer Experience

Lukas Schweiger and Jürg Meierhofer

Abstract Field service providers can reduce customer trips with evermore improving technologies in machine learning and digital twins. However, there is more than one aspect that needs to be considered. First, the customer's intimacy with the provider may shrink with reduced customer service journeys. Furthermore, the reduction of service trips may have a positive influence on ecological sustainability as CO_2 emissions are reduced. This paper presents an early-stage model to calculate three main KPIs in the service field: Ecological sustainability, Economic sustainability, and Customer intimacy. The model is built on an agent-based and discrete event core that simulates a service provider's service ecosystem. It is then tested on data from a real-life use case. The results are promising and show interrelationships that need to be further investigated in the future. A particular focus in further research has to lie in understanding the economic splitting of cost and benefit between the customer and the provider.

Keywords Process twin · Customer intimacy · Sustainable service · Service ecosystem simulation

1 Introduction

Service providers can improve their service offerings to their customers through evermore improving digital support in condition-based maintenance and service planning (Meierhofer et al., 2021). An example that got much attention during the last years of the pandemic is the possibility of remote service with a guided employee at the customer's location. These remote services will reduce service trips to the customer by car or other means of transport. However, with the help of a digital support system, it will be possible to prevent or reduce even those services that still need a trip

L. Schweiger (✉) · J. Meierhofer
Zurich University of Applied Sciences, CH 8401 Winterthur, Switzerland
e-mail: lukas.schweiger@zhaw.ch

J. Meierhofer
e-mail: juerg.meierhofer@zhaw.ch

J. Meierhofer et al. (eds.), *Smart Services Summit*, Progress in IS,
https://doi.org/10.1007/978-3-031-36698-7_22

to the customer. This digital support stems from improvements in machine learning for error and failure detection and in the field of digital process twins in industrial services (Singh et al., 2021) which reduce unnecessary trips. Machine learning for error and failure detection has become more affordable and relevant for SMEs in the last few years. Furthermore, digital process twins are changing the field by improving the understanding of service processes and helping service teams make short-term decisions. However, this change in technology leads to a change in the service ecosystem's function. Three significant aspects must be considered during this change phase, as illustrated in Fig. 1: 1. **Ecological sustainability**, 2. **Economic sustainability**, and 3. **Customer intimacy**. In this paper, we would like to explore the relationships between these three factors based on an agent-based model.

2 Background

In order to understand the triangle of tension between the ecological, economic, and customer intimacy goals of companies in the context of service operations in machine maintenance, we must first define what we mean by each of these terms. Nevertheless, firstly, we must also understand why this conflict is occurring and the technologies causing it. One of these technologies is the digital twin, as defined by Jones et al. (2020). Using such digital twins in the service ecosystem is one way of supporting the service organization in companies. At this point, however, we must specify more precisely what kind of digital twin we are talking about in this paper. According to Meierhofer et al. (2021), there are two types of digital twins: equipment twins, which are twins of machines or other equipment, and process twins, which digitally mirror service processes within and among companies. Therefore, we firmly believe that the digital process twin is causing the changes discussed in Sect. 1. This fact comes true, especially if the digital process twin is integrated into a broader system of digital support systems. We are working on such integration in an end-to-end process (Schweiger et al., 2022) for SMEs. Such a system will enable companies to improve their field service operations processes. However, we understand that conflicts of interest arise, such as the economic and ecological benefit of trips to

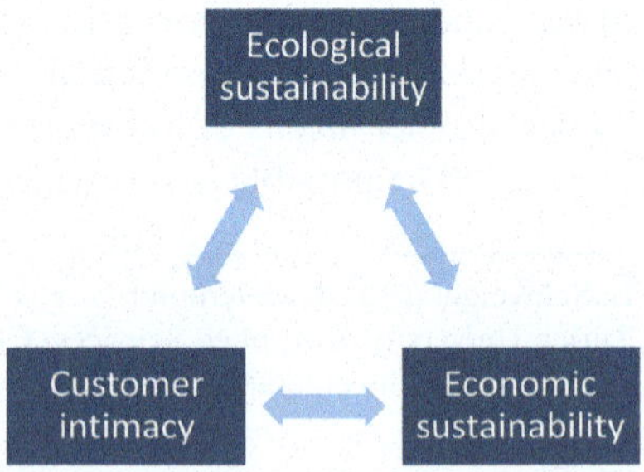

Fig. 1 Triangle of tension between: ecological sustainability, economic sustainability, and customer intimacy

customers avoided, however at the drawback of fewer physical customer interactions and thus reduced customer intimacy. This paper attempts to better understand these tensions with the help of a model and thus provide decision support for service organizations. For this purpose, we use a mixed model approach (Tjahjono & Jiang, 2015; Meierhofer et al., 2021) to model the service ecosystem. Below we describe the previously mentioned key performance indicators in detail.

2.1 Ecological Sustainability

In order to understand the influence of service trips on ecological sustainability, we consulted two different measurement methods. The first one is the Swiss concept of *Environmental Impact Points* (UBP'06) (Frischknecht & Büsser Köpfel, 2013). This UBP'06 method is of special interest to us as we mainly work together with companies from Switzerland. The method looks at three main factors: 1. Emissions, 2. Resources, and 3. Waste. The method splits emissions into the following four areas: 1. Air, 2. Noise, 3. Water, and 4. Soil. We are mainly interested in two aspects of the method for our research as we are looking into reducing trips in field service. For one, this is the emissions wherein we look at air pollution through CO_2; on the other hand, we are interested in the resources wasted by using fuel. The second method we consulted to build our model was the *Global Warming Potential* (GWP) which is built slightly differently but tackles the same problems concerning ecological sustainability. As we see the most significant influence on ecological sustainability in reducing car trips to the customers, we will use two KPIs from the two methods: 1. CO_2, and 2. Fuel consumption. Therefore we decided to go with the following two measurements for ecological sustainability: 1. $FuelConsumption = FC$, and 2. (CO_2).

The two measurement values are calculated in the model as a function of the distance driven d, and the type of the car, which in return defines the type of fuel used and influences the amount of (CO_2).

$$FC = f_{FC}(d[km], car[type]) \tag{1}$$

$$\mathrm{CO}_2 = f_{\mathrm{CO}_2}(FC, car[type]) \tag{2}$$

2.2 Economic Sustainability

Looking at the economic sustainability in service processes, we consulted the literature to understand better the economic impact of reduced service trips to the customer. This is especially important as already (Oliva & Kallenberg, 2003; Gebauer et al., 2005) have mentioned that it is hard to convince management of the benefits of improved service business. (Stenström et al., 2016; Wang et al., 2022; Sun et al., 2022) mention several influential parameters that need to be included. We generally look at cost factors C and benefit factors B when calculating the cost-benefit ratio (B/C). Stenström et al. (2016) compare corrective maintenance (CM) to preventive maintenance (PM). However, we are interested in the influence of predictive maintenance (PDM) on our three KPI measurements (Fig. 1). Therefore, we use the model of Stenström et al. (2016) to calculate the cost of predictive maintenance (C_{PDM}). Stenström et al. (2016), Wang et al. (2022) and Sun et al. (2022) use the same function to calculate the C_{PDM} as seen in Eq. 3. In our model, we use a selection of the influence factors they used in their calculations: 1. ProductionLoss = C_P, 2. Logistic Time = LT, 3. Repair Time = RT, 4. Cost of Material = C_M, and 5. Cost of Analysis = C_A adapted from C_I for a PDM case.

$$C_{PDM} = f_{C_{PDM}}(C_P, LT, RT, C_M, C_A) \tag{3}$$

In addition to those factors, we must consider the benefits the service trips generate for the provider. In our case, we look at the revenue generated through service contracts with the customer R_{SC} (Classen & Friedli, 2021), and the number of customers V_C (Stormi et al., 2018) that pay a service fee.

$$B_{PDM} = f_{B_{PDM}}(R_{SC}, V_C) \tag{4}$$

Therefore, in our model, we use the B/C ratio $\in \mathbb{R}\,\{0, 1\}$ as a KPI for the economic sustainability.

$$B/C = \frac{B_{PDM}}{C_{PDM}} \tag{5}$$

2.3 Customer Intimacy

Customer intimacy is the third element of the triangle of tension (Fig. 1). Customer intimacy is the "segmenting and tailoring offerings to precisely match the need of customers" according to Treacy and Wiersema (1993). In recent literature, there are different approaches on how to measure customer intimacy. We used two of them for our model. Habryn et al. (2010) uses a hierarchical approach that measures the Customer Intimacy Grade CIG, with the following influence factors: 1. Relationship Quality RQ, 2. Adaptability A, 3. Communication C, 4. Trust T, 5. Commitment C, 6. Relationship Bonds R, 7. Interaction I, 8. Activity Bonds A, and 9. Identification I.

The second method we used is the one from Bhatt and Sahil Bhanawat (2016), who see the following five factors influencing customer intimacy: 1. Reliability 2. Responsiveness 3. Assurance 4. Empathy 5. Tangibles. Based on those two approaches, we implemented the CIG calculation into our model, as displayed in Fig. 2. Customer

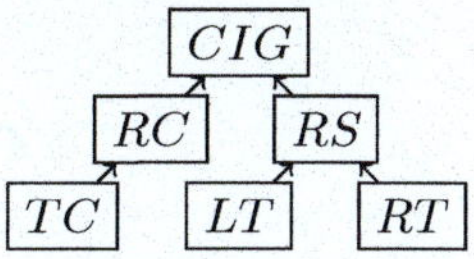

Fig. 2 CIG Decomposition adapted from Habryn et al. (2010) and Bhatt and Sahil (2016)

Intimacy Grade $CIG \in \mathbb{R}\{0, 1\}$ as a function of the Relationship Quality with Customer RC, which itself is a function of the tangibility of the service, which we measure in trips to the customer TC. Further, we look at the responsiveness of the service RS as a function of LT and RT. Therefore, the model will use the following formula to calculate the CIG:

$$CIG = f_{CIG}(TC, LT, RT) \tag{6}$$

3 Methodology

We use a simulation model of a service provider ecosystem to show the ecological and economic sustainability improvements based on an improved service process. The model is based on data from an industrial partner. For simplicity reasons, we reduced the model to its core. We use the mixed model approach (agent-based and discrete event) to simulate different actors in the service ecosystem. We use the following agents to simulate the ecosystem (see Fig. 3): 1. the service center Switzerland SC, 2. all service technicians in Switzerland ST, 3. all service vehicles in Switzerland SV, 4. all customers in Switzerland C, and 5. all machines installed in Switzerland M. Each of those agents is simulated in a Geographic information system (GIS) environment. This allows us to calculate and display the routes the ST have to travel with their SV if they fix a machine at a customer's site. The service process is modeled inside the service center as a discrete event process. This process evaluates if a service request sent by a customer requires an onsite intervention, i.e., a field trip of a service technician. Those field trips influence the ecological sustainability as each of them produces a certain amount of (CO_2). We calculate the (CO_2) production for each service trip per service vehicle SV_n in our simulation, as mentioned in Sect. 2.1, Eq. 2. Furthermore, we simulate the CIG for each customer C_n. As explained in Sect. 2.3, the CIG is dependent on the number of visits from the service technicians (Eq. 6). However, we average the CIG over all customers in the simulation. As mentioned in Sect. 2.2, we also look at the economic sustainability of the service using the cost-benefit ratio B/C. We measure the travel time LT, and the repair time RT of each service trip averaged over all service technicians and service incidents. Likewise, the cost of material C_M, and the cost of the production loss C_P are calculated for each service incident. Together with the cost of the analysis C_A, they build the cost factor for the B/C ratio in the simulation.

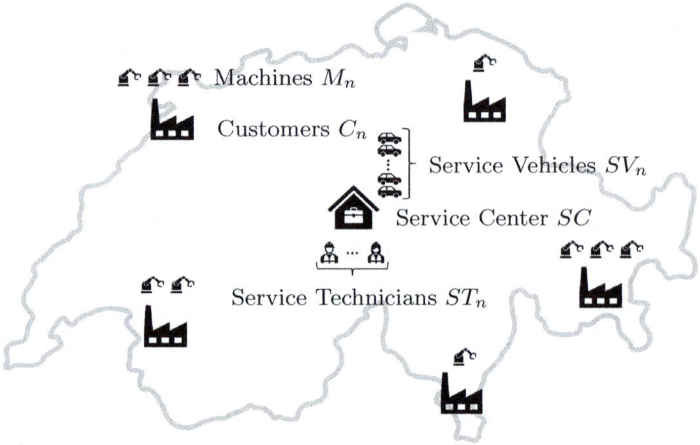

Fig. 3 Structure of simulation model

Based on this simulation structure, we calculated the following KPIs: 1. CIG, 2. B/C, and 3. (CO_2) by varying the input parameters: 1. Ratio of visits per service case r, 2. ST, and 3. SV. The ratio r is used as an indicator of how well the PDM is working. It influences the cost of the C_{PDM} and the probability that a service case leads to a field trip.

4 Findings

This first model can simulate the behavior of a service center with a variable number of ST and SV for field service. Furthermore, one can change the field trips per service case r. In the first run, the parameters ST and SV were set corresponding to the numbers of a real case study with an industrial partner in Switzerland.

The resulting KPI can be seen in Table 1 as a function of the ratio of onsite visits per service case, r. From this standpoint, one could say that a service is the

Table 1 Compare runs of model with actual ST and SV numbers of case

Run	#ST	#SV	r	CIG	B/C	CO_2
1	14	5	0.1	0.14	**0.91**	**0.754**
2	14	5	0.2	0.43	0.81	0.499
3	14	5	0.4	**0.86**	0.12	0.020
4	14	5	0.6	0.80	0.01	0.003
5	14	5	0.8	0.82	0.00	0.000
6	14	5	1.0	0.80	0.00	0.011

Table 2 Compare runs of model with $ST = SV$ numbers

Run	$\#ST$	$\#SV$	r	CIG	B/C	CO_2
1	5	5	0.1	0.14	**0.75**	**0.742**
2	5	5	0.2	0.40	0.47	0.475
3	5	5	0.4	0.80	0.06	0.037
4	5	5	0.6	**0.83**	0.03	0.044
5	5	5	0.8	0.82	0.01	0.007
6	5	5	1	0.81	0.00	0.00

most sustainable when the PDM model reliability works at its best (small r), and the trips to the customers are at a minimum. However, the level of r optimizing CIG is 0.4. In a variation of the experiment, all parameters were kept, but the number of technicians was reduced from 14 to 5. The hypothesis was that this would have the following impact: 1. (CO_2) unchanged, as this depends only on the service cases. 2. B/C reduced, as the fixed costs are reduced. 3. CIG reduced as the lower number of technicians results in longer waiting times and thus reduced responsiveness RS. The results shown in Table 2 confirm this hypothesis except for CIG. The customer intimacy RS in the numerical example chosen is considerably less sensitive to the responsiveness RS than expected.

Consequently, there is a need for further experiments and deeper investigation theoretical and on a practical level. A particular focus should lie on the improved understanding of the complex B/C ratio, which is modeled simplistically in this approach.

5 Conclusions and Recommendation

We connect three fields of interest in the service research community with our basic model. Thus, customer intimacy, economic goals, and ecological sustainability are integrated into one self-contained model. Through this, numerous insights were gained into the interconnection of the three different conceptual domains. Furthermore, the model lays the basis for further discussion about the compatibility of optimal customer support and achieving economic and ecological goals. This first approach represents just the beginning of further research in the field. Based on the promising results and feedback from industry partners, the understanding of the interplay between service process twins and their economic impact on service systems and the different actors in the respective ecosystems will be further developed. With this further research, we want to provide a tool to the industry to understand better their decisions in the design of their field service processes and their implications on their economic situation. A new economic layer will be added to the end-to-end process twin model (Schweiger et al., 2022). This layer should support decision-makers regarding economic or ecological decisions in service process twins.

References

Bhatt, A. K., & Sahil Bhanawat, D. (2016). Measuring Customer Satisfaction Using ServQual Model-An Empirical Study. *International Journal of Trend in Research and Development, 3*(1), 267–276.

Classen, M., & Friedli, T. (2021). It's not like we can charge for everything: Revenue models to capture value from smart services in Pacific Asia. *Asia Pacific Business Review, 27*(3), 405–430. https://doi.org/10.1080/13602381.2021.1894770

Frischknecht, R., & Büsser Köpfel, S. (2013). Swiss eco-factors 2013 according to the ecological scarcity method methodological fundamentals and their application in Switzerland. (Tech. Rep.). Bern.

Gebauer, H., Fleisch, E., & Friedli, T. (2005). Overcoming the service paradox in manufacturing companies. *European Management Journal, 23*(1), 14–26.

Habryn, F., Blau, B., Satzger, G., & Kölmel, B. (2010). Towards a model for measuring customer intimacy in B2B services. In *International conference on exploring services science* (pp. 1–14).

Jones, D., Snider, C., Nassehi, A., Yon, J., & Hicks, B. (2020). 5). Characterising the Digital Twin: A systematic literature review. *CIRP Journal of Manufacturing Science and Technology, 29*, 36–52. https://doi.org/10.1016/j.cirpj.2020.02.002

Meierhofer, J., Schweiger, L., Lu, J., Züst, S., West, S., Stoll, O., & Kiritsis, D. (2021). Digital twin-enabled decision support services in industrial ecosystems. *Applied Sciences (Switzerland), 11*(23). https://doi.org/10.3390/app112311418

Meierhofer, J., Schweiger, L., & Schreuder, L. (2021). Digital twin based decision support services in business operations. In *Smart services summit* (pp. 117–129).

Oliva, R., & Kallenberg, R. (2003). Managing the transition from products to services. *International Journal of Service Industry Management.*

Schweiger, L., Meierhofer, J., Barbieri, C., & Rapaccini, M. (2022). *End-to-End Digital Twin Approach for Near-Real-Time Decision Support Services* (p. 67). Smart Services Summit: Smart Services Supporting the New Normal.

Singh, M., Fuenmayor, E., Hinchy, E. P., Qiao, Y., Murray, N., & Devine, D. (2021). Digital twin: Origin to future. *Applied System Innovation, 4*(2). https://doi.org/10.3390/asi4020036

Stenström, C., Norrbin, P., Parida, A., & Kumar, U. (2016). Preventive and corrective maintenance-cost comparison and cost-benefit analysis Preventive and corrective maintenance-cost comparison and cost-benefit analysis. *Structure and Infrastructure Engineering, 12*(5), 603–617. https://doi.org/10.1080/15732479.2015.1032983

Stormi, K., Laine, T., Suomala, P., & Elomaa, T. (2018). Forecasting sales in industrial services. *Journal of Service Management, 29*(2), 277–300. https://doi.org/10.1108/JOSM-09-2016-0250

Sun, M., Ng, C. T.,Wu, F., & Cheng, T. C. (2022). Optimization of after-sales services with spare parts consumption and repairman travel. *International Journal of Production Economics, 244*, 108382. https://doi.org/10.1016/J.IJPE.2021.108382

Tjahjono, B., & Jiang, X. (2015). Linking symbiotic simulation to enterprise systems: Framework and applications. In Proceedings—winter simulation conference (pp. 823–834). Institute of Electrical and Electronics Engineers Inc. https://doi.org/10.1109/WSC.2015.7408219

Treacy, M., & Wiersema, F. (1993). Customer intimacy and other value disciplines. *Harvard Business Review, 71*(1), 84–93.

Wang, J., Wu, Y., Yang, Z., Chen, X., & Hu, W. (2022). Optimizing opportunistic preventive maintenance strategy for multi-unit system of CNC lathe. *Journal of Mechanical Science and Technology, 36*(1), 2022. https://doi.org/10.1007/s12206-021-1213-9

The Value of Guideline Practice as a Proto-Service for Smart Service Development

Petra Müller-Csernetzky, Mario Rapaccini, and Shaun West

Abstract In the context of digitalization, companies today are faced with the challenge of rethinking their process management, especially if they aim to provide smart services. Digitizing processes is not a direct route, as they are neither merely transferable nor do the processes will remain unchanged. The development and implementation of smart services as a dynamic form of digitalization also includes the consideration and analysis of existing workflows and practices. In particular, the embedded knowledge of interconnected and complex processes is underestimated, and digitization initiatives tend to move towards a closed end. In addition, the skills and embodied knowledge situated in employees is difficult to be used as core source for co-creative teamwork in front of transformation. Policy making and manual development used to support knowledge extraction and analysis of process management is a co-creative and highly effective support in this context. This paper sheds light on how the visual practice of guideline development from an industrial case can inform the creation of smart services in its primary state as proto service. This helps digital transformation project management become commercially attractive by reducing rework or project re-planning, and it enables employee integrity during the transformation. In addition, the developed approach can be used in such projects to support service training of employees in various roles from operations, development to management.

Keywords Facilitation · Transformation · Technical communication · Co-creation

P. Müller-Csernetzky (✉) · S. West
Lucerne University of Applied Sciences and Arts, Lucerne, Switzerland
e-mail: petra.mueller-csernetzky@hslu.ch

S. West
e-mail: shaun.west@hslu.ch

M. Rapaccini
Department of Industrial Engineering, University of Florence, Florence, Italy
e-mail: mario.rapaccini@unifi.it

J. Meierhofer et al. (eds.), *Smart Services Summit*, Progress in IS,
https://doi.org/10.1007/978-3-031-36698-7_23

1 Introduction

Today, digitization is a key challenge for manufacturing companies (Moon et al., 2018). Initiatives for digital innovation are in fact risky and unpredictable (Björkdahl, 2020). A typical situation concerns the introduction of Manufacturing Execution Systems (MES) to advance production planning, control, and optimization (Almada-Lobo, 2015). In this context it is crucial that the tacit knowledge possessed by blue collars, industrial engineers, and production managers, is not dispersed. Traditionally, this knowledge is explicated through the creation of vast repositories of technical procedures and user guidelines. More recently, the use of co-creative methodologies (Aarikka-Stenroos & Jaakkola, 2012) and visual facilitation, that are grounded on design principles, gained momentum (Freytag & Storvang, 2016). Participatory Action Research (PAR) (Kemmis et al., 2014) is used to explore how high-tech environments facing the challenges of digitalization can benefit from this approach. The study outlines that facilitation through policy development practice makes a significant contribution to a sustainable and people-centered digital transformation of organizations.

2 Theoretical Background

Challenges of knowledge management are discussed widely associated to practices of IT management (Teece, 1981; Tsoukas, 2009), particularly in knowledge-intensive sectors, such as high-tech industries (Björkdahl, 2020). In these contexts, novel knowledge is generated and shared through technical drawings and operational procedures (Hughes, 2002). The sharing of existing and codified knowledge as well as the development of novel one is challenging, especially where collaboration is hindered by the typical silo mentality of engineering departments (Nonaka & Teece, 2001). Although knowledge is created from information (Rowley, 2007), managing knowledge in complex organizations requires practices, such as the well-known SECI-spiral model (Nonaka, 1994; Nonaka et al., 2006) (see Fig. 1), that are different from those used for information management (Bratianu, 2015).

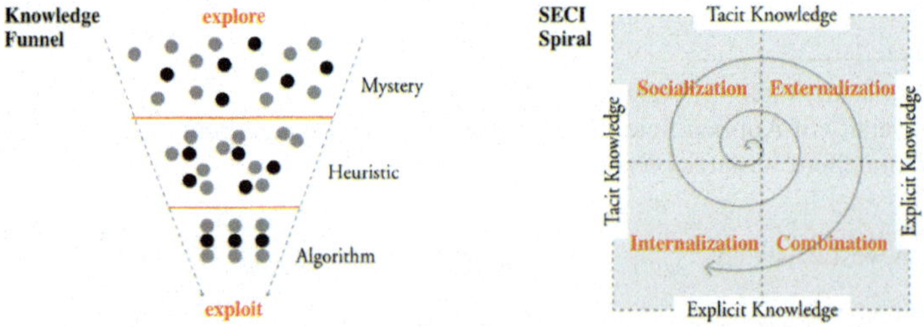

Fig. 1 The knowledge funnel (Martin, 2009) and the SECI-spiral (Nonaka, 1994)

The use of visual techniques to aid generation of individual and organizational knowledge is rather new but widely acknowledged (Rill, 2016), especially in product, process, and business model innovation (Martin, 2009; Täuscher & Abdelkafi, 2017). In these contexts, knowledge can be generated iteratively (see the Martin's knowledge funnel of Fig. 1) when generating solutions from ill-structured problems, having a high degree of ambiguity. Methods developed from innovation management such as storytelling to express individual experiences (Perret et al., 2004; Szkupinski-Quiroga, 2015), or life drawing (Varanka & Usery, 2018) are further examples of how visual approaches support the informative expression of tacit knowledge. To understand these mechanisms, the conceptualization of boundary objects can be helpful. Boundary objects, such as documents, notes, illustrations, and drawings, are information-laden objects that serve to clarify the different perspectives and the commonalities between the borders of individual perception (Carlile, 2002). In this sense, they either can help the co-ordination or can uncover inconsistencies between parties and their communication (Star, 2010). Therefore, any visual representation can be viewed as a boundary object, as far as it provokes reflections and/or eases interactions, based on the way it presents a certain topic (Singh, 2011; Whyte et al., 2008). Facilitators use the potential of visualization to help participants in a workshop to see aspects more clearly (Brown, 2019; Täuscher & Abdelkafi, 2017). In change management, it is used in the settlement of new and lasting routines (Turner, 1982) and to be aware of the pitfalls along the transformation (Errida & Lofti, 2021; Kotter, 2007). Co-creational approaches are widely used in design activities (Steen, 2013) to engage employees (Aarikka-Stenroos & Jaakkola, 2012) and foster trust and relationship (Freytag & Storvang, 2016; Tann, 2021). Last, the idea of seeing visual practices as proto services derives from the notion that value develops through co-creational and contextual application of specialized competencies, as described by the premises of the service-dominant logic (Vargo & Lusch, 2014; Vargo et al., 2010). The logic of value creation is then determined by the surrounding system, its interactional characteristics, and its resources in place (Lusch & Vargo, 2012). Within the study, especially the aspects, that all actors in the project are resource integrators as part of the complex information network they support the forthcoming by actively contributing interactionally their knowledge and skills (Vargo & Lusch, 2008).

3 Methodology

The investigation into a long-term case study at a hi-tech company provided the possibility to apply Participatory Action Research (PAR) (Kemmis et al., 2014). Practical approaches of technical and visual communication facilitated the policy making for implementing a new MES. Within two and a half years, three guidelines were developed, and its development process builds the source of the material for analysis. A document analysis, interviews and recordings of work sessions helped to understand the practice and the impact within the transformation initiative. The rather quantitative results from the document analysis underpin results from the qualitative approach from the coding of material such as memos, interviews, and meeting recordings.

The in-vivo material coming from the researcher is analyzed using the multilevel-interpretation approach (Alvesson & Sköldberg, 2018), especially since the material contains meaning in different ways. As a supportive structure for understanding behavioral aspects, the Activity System by Engeström (1987), helped to understand practices in the complex project context. Aspects such as roles, the division of labor, the community, and its mediating artefacts as well as the subject-object-outcome relationship are explored. The document analysis (Bowen, 2009) was applied to all the iterations of the guidelines and their page content (text types, diagrams, screen-shots, photography, flow charts etc.). It can be measured to provide information about the changing context of guideline use. Codes are identified by their visual category, relation, origin, quality, and their meaning. Hence, the page contents of all guidelines had been coded, measured, and collected. This approach was applied at start, in the middle and at the end of the development to compare the document stages, considering variations in between.

Since the investigation takes place at an industrial site, the PAR, as a qualitative research methodology, strives to understand challenges between people and act to effect change. As the challenges lie in the communication between the employees, it has been beneficial to involve the participants in every step of the development (Burian et al., 2010). Since the development of the software in the organization inevitably lead to successive definable work phases, the work process in the guideline development became cyclical as well. The team members understood the co-creative and cyclical course of activities as very supportive and guideline fragments were jointly observed and evaluated by researcher and participant leading to follow-up actions (Kapoor, 2002; Riel & Polin, 2004). Here, the focus lies on social action and their transformation for the better by dissolving barriers of communication. This approach is advocated by many, since through its integration and application into everyday life, the method may motivate the transformation of the organization from the bottom up (Berg, 2004). As the participants showed interest in the improvement of issues, they became critical and self-directed, playing a more responsive part in the process.

4 Results of Case Study

There were three Action Research cycles. Coding supported the analysis of the meeting and interview records as well as the memos. It allowed for an understanding of participants' perceptions of changes in guideline content as well as interaction throughout the project. The application of the questions from the Activity System provided insights into how the change project was handled internally. It made the interaction of actions more transparent and the position of the project in the complex organizational network. The document analysis helped to understand the different qualities of the content and the dynamic information architecture based on different page types.

The **first cycle** focused on finding a process to deal with different visual practices and mind sets. Along the way, the project offered a good basis for investigating how visual practices can support the development of a visual template system for the MES application's interface for the production employees. It became apparent that visualization and iterative reflection form a process that help the engineers to express and trigger a visual culture. The guideline expert and the templates were supportive for employees, when the use was explained slowly, iteratively, and visually.

The **second cycle** dealt with the overarching information complexity, and especially with the use of the guidelines in the discussions. The increase in work complexity in the project allowed for a focused search for an information architecture differentiation between producing, planning, and managing knowledge. The scope of work led to a change in the perception of user manuals as technical documents only and moving away from guidelines that purely document templates. The utilized guideline fragments became a strategic tool, and it enabled the integration of a discussion culture in which transformation can be actively experienced.

The **third cycle** focused on the way of dealing with professionals and develop principles and scalability. After the adjustment in the software, two guidelines were almost completely revised. The user groups were redefined and three guidelines for different levels of work with and about the application were agreed upon. This cycle created insight into different levels of competencies depending on the role. The facilitation needed to focus on delustering complexity by a co-creative, visual, reflective, and personal process. The scope of the guideline delivery was juggled so that the value lied in the development of principles which made the transformation manageable.

5 Discussion

Organizations look for support when certain skills are not present and hinder forthcoming. In the study, the challenges arose from difficulties related to a lack of meta-skills (Choudrie & Selamat, 2006) and a low flexibility in verbal and visual communication culture. The guideline expert was brought in to reduce the complexity of communicating about the use of a new MES and to clearly communicate the benefits of the different work levels in the system. In addition, the expert was hired while the MES implementation was ongoing, and tasks were changing during the project. The assignment changed from a point-by-point support to a process-oriented flow of support that served the work analysis before the actual application was in use.

The idea of using a visual approach for analyzing work routines in the context of highly technological environment is promising, and at the same time very challenging due to a lack of an active visual sharing culture in hi-tech industry. The role of the guideline expert in the project has been challenged to constant flexible and deliberately used facilitation (Tann, 2021) using guideline practice supporting the project's forthcoming. The role exceeds the task of pure content reproduction, as a joint iterative problem-solving process (Aarikka-Stenroos & Jaakkola, 2012) structures the interaction between expert and employees. The challenges in the study are twofold. The first challenge is, that the facilitation supports the re-processing of routines and subsequently the re-definition of them as newly set routines in the MES. The second practical challenge is, to bring about the development and organization of the information architecture of the guidelines, which are to be used by new employees.

The methodical testing of guiding fragments facilitates the preliminary work for digitization. There are many barriers to sharing knowledge, and interaction promotes understanding of complex issues (Dreyer & Bown, 2017). Guideline-oriented facilitation is an empathic and respectful way to work through this process together in a one-to-one interaction to externalize knowledge (Tsoukas, 2009). It was also found that iterative human-based facilitation (Hu, 2016) promotes looped learning to make better decisions about information structures. The study showed an improved organizational effectiveness of empowerment through one-to-one interactions as facilitation (Tann, 2021) lead to co-creative joint problem solving (Aarikka-Stenroos & Jaakkola, 2012) (Fig. 2). Through teamwork and open reflection, valuable solutions are found through the process itself. All actors in the project are resource integrators (Vargo & Lusch, 2014) as part of the complex information network contributing interactionally knowledge and skills. Developing principles for using and implementing the new system supported staff learning to solve future challenges (Kotter, 2007; Muller & Zenker, 2001).

In short, the different knowledge types had to be differentiated by targeted compositions of content. The information architecture grew by elaborating the different roles and knowledge types. This process involves the reorganization of the system based on the team's findings. The change from analog to digital starts by using guideline fragments and provide value when the content is clarified through iterations. When the fragments went through this co-creative and re-learning process,

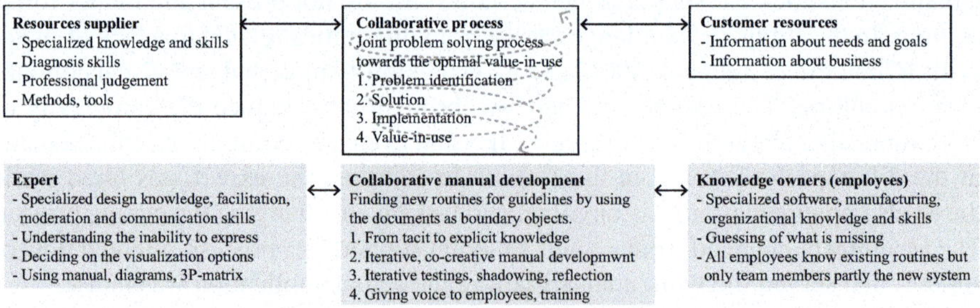

Fig. 2 Applied joint value co-creation in gray (Aarikka-Stenroos & Jaakkola, 2012)

they form final documents. The approach consists of the following steps building an *action framework*:

1. Listening, recording and clarification of actions
2. Sorting and listing the different kinds of activities
3. Defining and coding content by variants of visual/text combinations
4. Coordinating the roles to the activities by evaluation
5. Assembling according to role and combining fragments
6. Defining complete content of guideline and release.

Limitations

This kind of value creation is possible in a working relationship based on trust. Like the handling of the documents as they stayed at the expert and not at the organization. When files are shared from start, their strategic character is influenced by others.

6 Implications

The study offers an action framework for the facilitation of work process analyses. It forms the basis for a system that can address very complex challenges and provide a comprehensive contextual understanding of how to deal with knowledge and information architectures. The process itself only takes effect in the organization and in defined spaces of knowledge within the team. The framework creates principles for an organizational learning culture that enables the organization to manage transformations of varying scale and scope. It involves employees from production to planning to management to promote long-term transformation and sustainable growth. Therefore, this approach is a people-based service that can serve as a precursor to further implementations of information architectures. Be it for the development of smart services and the associated training of employees or to enable further stages of digitization.

References

Aarikka-Stenroos, L., & Jaakkola, E. (2012). Value co-creation in knowledge intensive business services. *Industrial Marketing Management , 41*, 15–26.

Almada-Lobo, F. (2015). The industry 4.0 revolution and the future of Manufacturing Execution Systems (MES). *Journal of Innovation Management, 3*(4), 16–21.

Alvesson, M., & Sköldberg, K. (2018). *Reflexive methodology—New vistas for qualitative research* (3rd ed.). Sage.

Berg, B. L. (2004). *Qualitative reserach methods for the social sciences* (5th ed.). Pearson.

Björkdahl, J. (2020). Strategies for digitalization in manufacturing firms. *California Management Review, 62*(4), 17–36.

Bowen, G. (2009, August). Document analysis as a qualitative research method. *Qualitative Research Journal, 9*(2), 27–40.

Bratianu, C. (2015). *Organizational knowledge dynamics: Managing knowledge creation, acquisition, sharing, and transformation.* IGI Global.
Brown, T. (2019). *Change by design.* HarperCollins Publishers.
Burian, P. E., Rogerson, L., & Maffei III, F. R. (2010, August). The research roadmap. *Contemporary Issues In Education Research, 3*(8), 43–57.
Carlile, P. R. (2002, July–August). A pragmatic view of knowledge and boundaries: Boundary objects in new product development. *Organizational Science, 13*(4), 442–455.
Choudrie, J., & Selamat, M. H. (2006). The consideration of meta-abilities in tacit knowledge externalization and organizational learning. In *Proceedings of the 39th Hawaii international conference on system sciences.*
Dreyer, H., & Bown, R. (2017). Tacit knowledge triggers. In *The 1st international conference on family business and entrepreneurship.*
Engeström, Y. (1987). *Learning by expanding.* Orienta-Konsultit.
Errida, A., & Lofti, B. (2021). The determinants of organizational change management success. *International Journal of Engineering Business Management, 13,* 1–15.
Freytag, P. V., & Storvang, P. (2016). Dynamics of a facilitator's role: Insights from the Danish construction industry. *Management Revue, 27*(3), 117–138.
Hu, Y. (2016). The opportunities and limits of organizational learning. In *European conference on knowledge management* (pp. 1139–1142). Kidmore End.
Hughes, M. (2002, August). Moving from information transfer to knowledge creation. *Technical Communication, 49*(3), 275–285.
Kapoor, I. (2002, February). The Devil's in the theory: A critical assessment of Robert chambers' work on participatory development. *3rd World Quarterly, 23*(1), 101–117.
Kemmis, S., McTaggart, R., & Nixon, R. (2014). *The action research planner.* Springer.
Kotter, J. P. (2007, May–June). Leading change: Why transformation efforts fail. *HBR.*
Lusch, R. F., & Vargo, S. L. (2012). The forum on markets and marketing (FMM): Advancing service-dominant logic. *Marketing Theory, 12*(2), 193–199.
Martin, R. (2009). *The design of business.* Harvard Business Review Press.
Moon, I., Lee, G. M., Park, J., Kiritsis, D., & von Cieminski, G. (2018). *Advances in production management systems. Smart manufacturing for industry 4.0.* Springer Nature.
Muller, E., & Zenker, A. (2001). Business services as actors of knowledge transformation. *Research Policy, 30,* 1501–1516.
Nonaka, I. (1994, February). A dynamic theory of organizational knowledge creation. *Organization Science, 5*(1).
Nonaka, I., & Teece, D. (2001). *Management industrial knowledge.* In I. Nonaka & D. Teece (Eds.). Sage Publications.
Nonaka, I., von Krogh, G., & Voelpel, S. (2006). Organizational knowledge creation theory: Evolutionary paths and future advances. *Organization Studies, 27*(8), 1179–1208.
Perret, R., Borges, M. R., & Santoro, F. M. (2004). *Applying group storytelling in knowledge management.*
Riel, M., & Polin, L. (2004). Learning communities: Common ground and critical differences in designing technical support. In S. Barab, R. Kling, & J. Gray (Eds.), *Designing for virtual communities in the service of learning.* Cambridge University Press.
Rill, B. (2016). Resonant co-creation as an approach to strategic innovation. *Journal of Organizational Change Management, 29*(7), 1135–1152.
Rowley, J. (2007, February). The wisdom hierarchy: Representations of the DIKW hierarchy. *Journal of Information Science, 33*(2), 163–180.
Singh, A. (2011). Visual Artefacts as boundary objects in participatory research paradigm. *Journal of Visual Art Practice, 10*(1), 35–50.
Star, S. (2010). This is not a boundary object: Reflections on the origin of a concept. *Science, Technology, & Human Values, 35*(5), 601–617.
Steen, M. (2013, Spring). Co-design as a process of joint inquiry and imagination. *Design Issues, 29*(2), 16–28.

Szkupinski-Quiroga, S. (2015). Performative storytelling an knowledge creation.
Täuscher, K., & Abdelkafi, N. (2017). Visuel tools for business modell innovation. *Creativity and Innovation Management, 26*(2).
Tann, J. (2021). The change agent in innovation. *Prometheus, 37*(1), 44–53.
Teece, D. J. (1981). The market for know-how and the efficient international transfer of technology. *The Annals of the American Academy of Political & Social Science, 458*, 81–96.
Tsoukas, H. (2009, November–December). A dialogical approach to the creation of new knowledge in organizations. *Organization Science, 20*(6), 941–957.
Turner, A. N. (1982, September). *Harvard Business Review*. Retrieved July 2020, from Consulting Is More Than Giving Advice www.hbr.com
Varanka, D. E., & Usery, L. E. (2018). The map as knowledge base. *International Journal of Cartography*.
Vargo, S. L., & Lusch, R. F. (2008). Service-dominant logic: Continuing the evolution. *Journal of the Academic Marketing Science, 36*, 1–10.
Vargo, S. L., & Lusch, R. F. (2014). Inversions of service-dominant logic. *Marketing Theory, 14*(3), 239–248.
Vargo, S. L., Lusch, R. F., & Akaka, M. A. (2010). Advancing service science with service-dominant logic—Clarifications and conceptual developments. In P. P. Maglio, C. A. Kieliszewski, & J. C. Spohrer (Eds.), *Handbook of service science* (pp. 133–156). Springer.
Whyte, J., Ewenstein, B., Hales, M., & Tidd, J. (2008, February). Visualizing knowledge in project-based work. *Long Range Planning*, 74–92

The manufacturer's authorised representative in the EU is Springer Nature Customer Service Centre GmbH, Europaplatz 3, 69115 Heidelberg, Germany. If you have any concerns regarding our products, please contact ProductSafety@springernature.com

Printed and bound by CPI Group (UK) Ltd, Croydon, CR0 4YY
01/10/2026
02225902-0004